ESTIMATION OF
ORGANIC COMPOUNDS

T0296960

ESTIMATION OF
ORGANIC COMPOUNDS

BY

F. WILD
M.A., PH.D., F.R.I.C.

*Fellow and Senior Tutor of Downing College, and
Research Chemist, Department of Medicine,
University of Cambridge*

CAMBRIDGE
AT THE UNIVERSITY PRESS
1953

CAMBRIDGE UNIVERSITY PRESS
Cambridge, New York, Melbourne, Madrid, Cape Town, Singapore, São Paulo, Delhi

Cambridge University Press
The Edinburgh Building, Cambridge CB2 8RU, UK

Published in the United States of America by Cambridge University Press, New York

www.cambridge.org
Information on this title: www.cambridge.org/9780521112758

First published 1953
This digitally printed version 2009

A catalogue record for this publication is available from the British Library

ISBN 978-0-521-06800-0 hardback
ISBN 978-0-521-11275-8 paperback

CONTENTS

PREFACE

The constitution of an organic compound is established by a complete analysis of its elements and by characterization of its groupings. In simple molecules it may only be necessary to find the percentage of carbon and hydrogen and identify a single grouping, but with more complex structures each of the characteristic groupings should be estimated. The principles of a limited number of analytical methods are described in textbooks of organic chemistry, but the improved laboratory techniques which are now available are hardly ever mentioned and even more rarely used. Many of the newer simple methods remain unnoticed and untried, although often they would give better results.

The estimation of groupings is rarely dealt with adequately in the teaching of organic chemistry as little or no time is set aside in laboratory courses for quantitative analysis, although its fundamental importance in the determination of structure is stressed in the lecture room. A few of the simpler experiments for the determination of the commoner groupings should be included at an early stage in all elementary practical courses, not only to provide a link with the lecture room and the textbook, but also to introduce into laboratory work an accuracy which is frequently absent from preparative organic chemistry. A wider range of methods should be established in advanced courses, to provide the future research worker with the necessary fundamental knowledge of organic analysis.

I hope that this book will help to stimulate interest in this branch of chemistry, by presenting methods which are used for the estimation of organic compounds. Selected general methods of analysis and the estimation of a number of specific simple substances are described in detail. The specialized methods—often involving individual techniques—used for amino acids, the sugars and the sulphonamides are not included.

I wish to express my gratitude to my wife for her constant help and advice in the preparation of the manuscript, to Mr A. R. Gilson for his assistance in the section on hydrogenation and in particular for providing the sketches for figs. 1, 2, and 3, and to Messrs British Drug Houses Limited and Messrs Townsend and Mercer Limited for permission to describe and reproduce in fig. 11 a simplified and improved form of apparatus for the Karl Fischer reagent. I also wish to thank the staff of the University Press for their patient collaboration, without which this volume would have been impossible.

F. W.

CAMBRIDGE
10 *October* 1952

CHAPTER I

OLEFINES

SOLID AND LIQUID OLEFINES

The characteristic property of olefines, their combining directly with certain atoms or groups, is used for the estimation of unsaturated compounds. The reagents and the experimental conditions must be controlled carefully to make the addition quantitative and to prevent side reactions occurring. The rate of addition varies considerably with the types of grouping attached to the olefinic linkage and also with the degree of substitution. The following reactions have been used to measure the degree of unsaturation.

Hydrogenation:

$$\text{>C=C<} + 2H = \text{>CH—CH<}$$

   A. Macro-hydrogenation.
   B. Micro-hydrogenation.

Bromine number:

$$\text{>C=C<} + Br_2 = \text{>C—C<} \quad \text{Br Br}$$

   A. Direct titration.
   B. Electrometric titration.
   C. Bromine absorption.

Iodine number:

$$\text{>C=C<} + I_2 = \text{>C—C<} \quad \text{I I}$$

   A. Wijs' method.
   B. Hanus' method.
   C. Rosenmund-Kuhnhenn's method.

Thiocyanogen number:

$$\text{>C=C<} + (SCN)_2 = \text{>C—C<} \quad \text{NCS SCN}$$

Estimation by peracids:

$$\text{>C=C<} + R.CO.O.OH \longrightarrow \text{>C}\underset{O}{\text{——}}\text{C<}$$

Miscellaneous:

HYDROGENATION

$$\text{>C=C<} + 2H = \text{>CH—CH<}$$

Catalytic hydrogenation is used extensively to determine the degree of unsaturation in organic compounds by measuring the volume of hydrogen absorbed by a weighed amount of the substance. The method is accurate, and if necessary it can be done on a micro-scale. The experimental conditions, such as temperature, hydrogenation pressure, catalyst, solvent and purity of the compound, are all interdependent, so that it is impossible to give precise details for the reduction of a particular type of structure. Olefines normally are saturated under mild conditions, and it is convenient to carry out the reaction at atmospheric pressure.

The apparatus is a hydrogen reservoir connected to a reaction vessel which can be shaken mechanically and heated if necessary. The system is filled with hydrogen, equilibrium is reached between the catalyst and hydrogen, a weighed amount of the substance is introduced into the reaction vessel, and the hydrogen absorbed on shaking, is measured either by a reduction in the volume or by a fall in the pressure. Numerous forms of apparatus have been described by which these changes can be followed in the laboratory.[1]

The types of micro-apparatus so far constructed have been based on the principle of the Barcroft differential manometer.[2] Two similar flasks are connected to a hydrogen reservoir and to

[1] See C. Paal and J. Gerum, *Ber.* 1908, **41**, 805–17; R. Willstätter and D. Hatt, *ibid.* 1912, **45**, 71–81; A. Skita and W. A. Meyer, *ibid.* 1912, **45**, 3589–95; O. Stark, *ibid.* 1913, **46**, 2335–6; and for pressures up to 3·5 atm. see R. Adams and V. Voorhees, *Org. Syn.* 8, 10–16, col. vol. 1st ed. pp. 53–9, 2nd ed. pp. 61–7; for pressures up to 300 atm. and temperatures up to 400° C. see H. Adkins, *J. Amer. Chem. Soc.* 1933, **55**, 4272–9; for pressures up to 300 atm. and temperatures up to 250° C. see A. R. Gilson and T. W. Baskerville, *Chem. and Ind.* 1943, **62**, 450–2 (fig. 1).

[2] J. Barcroft, *J. Physiol.* 1908, **37**, 212.

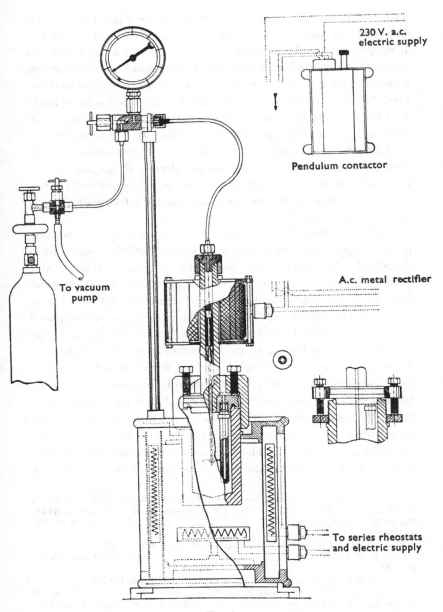

To vacuum pump

230 V. a.c. electric supply

Pendulum contactor

A.c. metal rectifier

To series rheostats and electric supply

Fig. 1.

4 OLEFINES

each other through a sensitive capillary tube filled with dibutyl phthalate or clove oil. In most forms of the apparatus the hydrogenation is done in one flask and solvent is added to the other which is used as a compensating vessel,[1] although sometimes it is used for the reduction of a control substance and the final difference in the volume of hydrogen absorbed is measured.[2] The latter method is used frequently but requires very rigid thermostatic control for accurate results. It has been used to compensate empirically for the incomplete reduction of substances, such as the polyenes.[3] In all micro-hydrogenations it is essential that each piece of apparatus should be perfectly clean and all chemicals pure, as traces of impurity frequently cause inconsistent results.[4]

A wide variety of catalysts, including platinum, palladium, nickel, copper chromite and a number of oxides, have been described. Platinum and palladium in the colloidal and amorphous forms, and less often Raney nickel, are used most frequently for the quantitative reduction of double bonds. A very active platinum catalyst (Adams' catalyst) is prepared by reducing the hydrated platinum oxide (PtO_2, H_2O) by shaking with hydrogen at room temperature. The dioxide is formed by fusion of chloroplatinic acid with sodium, or potassium, nitrite at 500–550° C.[5] The chloroplatinic acid is recovered by dissolving the residue in aqua regia, but if the catalyst has been poisoned extensively it should be purified by making ammonium chloroplatinate and using this as the starting material for the fusion.[6] The fusion can be carried out successfully in a beaker, but some skill is needed to produce a very

[1] J. H. C. Smith, J. Biol. Chem. 1932, 96, 35–51; H. Jackson and R. N. Jones, J. Chem. Soc. 1936, pp. 895–9; H. Jackson, J. Soc. Chem. Ind. 1938, 57, 1076–7; J. F. Hyde and H. W. Scherp, J. Amer. Chem. Soc. 1930, 52, 3359–63.
[2] R. Kuhn and E. E. Möller, Angew. Chem. 1934, 47, 145–9.
[3] Cf. R. F. Milton and W. A. Waters, Methods of Quantitative Microanalysis, pp. 568–9.
[4] A. E. Gillam and M. S. El Ridi, Biochem. J. 1936, 30, 1735–42.
[5] R. Adams, V. Voorhees and R. L. Shriner, Org. Syn. 8, 92–9, col. vol. 1st ed. pp. 452–60, 2nd ed. pp. 463–70; A. H. Cook and R. P. Linstead, J. Chem. Soc. 1934, pp. 946–56.
[6] E. L. Baldesweiler and L. A. Mikeska, J. Amer. Chem. Soc. 1935, 57, 977–83; W. F. Bruce, ibid. 1936, 58, 687–8 and Org. Syn. 17, 98.

active catalyst. A fusion apparatus has been described by Short.[1] Platinum black can be used either in the amorphous form or supported on an inert carrier. The former is prepared by reducing chloroplatinic acid with formaldehyde in the presence of sodium hydroxide,[2] and the latter by warming a suspension of the inert material in chloroplatinic acid, precipitating platinum hydroxide on the carrier and reducing. The carriers used most frequently are barium sulphate, charcoal, calcium carbonate, kieselguhr and silica gel.[3] Palladium catalysts are prepared by methods similar to those used for platinum. Soluble salts can be reduced with formaldehyde in the presence of sodium hydroxide;[4] the oxide, prepared by fusion of the chloride with sodium nitrite, can be reduced with hydrogen at room temperature;[5] and palladium black can be deposited on an inert carrier as, for instance, barium sulphate,[6] activated charcoal,[7] calcium or strontium carbonate,[8] silica gel[9] or kieselguhr.[10] Until recently the nickel catalyst most commonly used in the laboratory was prepared by impregnating a porous support with a nickel salt, forming the oxide and finally reducing with hydrogen at 300–400° C. This form of the catalyst could not be used at atmospheric pressure and room temperature, and it was not possible to use nickel under these conditions until 1927 when

[1] W. F. Short, J. Soc. Chem. Ind. 1936, 55, 14 T.

[2] R. Feulgen, Ber. 1921, 54, 560–1; O. Loew, ibid. 1890, 23, 289–90; R. Willstätter and D. Hatt. loc. cit. (p. 2, n. 1); R. Willstätter and O. Waldschmidt-Leitz, ibid. 1921, 54, 113–38.

[3] Houben-Weyl, Die Methoden der organischen Chemie, 2, 499; H. Kaffer, Ber. 1924, 57, 1261–5; N. D. Zelinsky and M. B. Turowa-Pollak, ibid. 1925, 58, 1298–1303; M. Latshaw and L. H. Reyerson, J. Amer. Chem. Soc. 1925, 47, 610–12; V. N. Morris and L. H. Reyerson, J. Phys. Chem. 1927, 31, 1220–9; ibid. 1927, 31, 1332–7.

[4] Houben-Weyl, Die Methoden der organischen Chemie, 2, 498.

[5] R. L. Shriner and R. Adams, J. Amer. Chem. Soc. 1924, 46, 1683; cf. R. Adams, V. Voorhees and R. L. Shriner, loc. cit. (p. 4, n. 5); A. H. Cook and R. P. Linstead, loc. cit. (p. 4, n. 5); W. F. Bruce, loc. cit. (p. 4, n. 6).

[6] E. Schmidt, Ber. 1919, 52, 400–13; R. L. Shriner and R. Adams, loc. cit. (p. 5, n. 5).

[7] W. H. Hartung, J. Amer. Chem. Soc. 1928, 50, 3370–4; E. Ott and R. Schröter, Ber. 1927, 60, 786–804; F. Mayer and G. Stamm, ibid. 1923, 56, 1424–33; C. Mannich and E. Thiele, Ber. deut. pharm. Ges. 1916, 26, 36–48; J. Soc. Chem. Ind. 1916, 35, 548–9.

[8] M. Busch and H. Stöve, Ber. 1916, 49, 1063–71.

[9] M. Latshaw and L. H. Reyerson, loc. cit. (p. 5, n. 3).

[10] Th. Sabalitschka and W. Moses, Ber. 1927, 60, 800–1.

Raney prepared a catalyst of greatly increased activity. An improved method of preparation has been described recently which gives a highly active catalyst ('W-6') which can be used for hydrogenation at low pressures.[1] A nickel-aluminium alloy, corresponding to $NiAl_2$, is prepared and the aluminium is dissolved out with sodium hydroxide.[2] The residual nickel is pyrophoric and is kept under water or under a suitable organic liquid. The amount of this catalyst required for a given hydrogenation is many times greater than for a catalyst of nickel on kieselguhr.

The activity of catalysts frequently shows a marked response to the presence of traces of promoters, and, indeed, the most successful catalysts are often very impure. Adams' catalyst contains sodium salts and Raney nickel contains aluminium and usually cobalt, and part of their activity probably arises from these impurities. An increase in activity may be gained by adding promoters. The activity of Raney nickel is increased greatly by the addition of small amounts of metals such as platinum, copper, zinc, chromium, molybdenum, iron and caesium;[3] small amounts of mineral acids increase the rate of hydrogenation of benzene using platinum black,[4] and benzoyl peroxide frequently increases the velocity of hydrogenation.[5] The efficiency of the catalysts can be reduced partially or completely by poisons. The former is sometimes done deliberately to stop the reduction at an intermediate stage, as in the hydrogenation of an acid chloride to an aldehyde,[6] but the latter arises from impurities or from the reaction. Mercury is described as a poison, although small quantities of the metal can be added to the reaction vessel and seem to have little effect when palladium

[1] H. R. Billica and H. Adkins, *Org. Syn.* **29**, 24–9.

[2] M. Raney, U.S. pat. 1,628,190; H. Adkins and H. R. Billica, *J. Amer. Chem. Soc.* 1948, **70**, 695–8; L. W. Covert and H. Adkins, *ibid.* 1932, **54**, 4116–17; R. Mozingo, *Org. Syn.* **21**, 15–17.

[3] J. R. Reasenberg, E. Lieber and G. B. L. Smith, *J. Amer. Chem. Soc.* 1939, **61**, 384–7; M. Delépine and A. Horean, *Compt. rend.* 1935, **200**, 1414.

[4] K. Kindler, E. Brandt and E. Gehlhaar, *Ann.* 1924, **511**, 209–12; K. Kindler and W. Peschke, *ibid.* 1935, **519**, 291–6; J. H. Brown, H. W. Durand and C. S. Marvel, *J. Amer. Chem. Soc.* 1936, **58**, 1594–6.

[5] G. Thomson, *ibid.* 1934, **56**, 2744–7.

[6] K. W. Rosenmund and F. Zetzsche, *Ber.* 1921, **54**, 425–40; K. W. Rosenmund, F. Zetsche and G. Weiler, *ibid.* 1923, **56**, 1481–7.

charcoal is the catalyst.[1] Unsaturated compounds prepared by dehydration with thionyl chloride usually resist hydrogenation, as they still retain traces of sulphur products. These can be removed generally by shaking with aluminium amalgam in moist ether.[2] Other sulphur compounds, including vulcanized rubber, have been described as poisons, and it is recommended that any rubber used in a hydrogenation apparatus should be freed from sulphur as far as possible by boiling with sodium hydroxide and water.[3] It has been suggested that the poisoning elements, such as sulphur, selenium, tellurium, phosphorus and arsenic, are ineffective as poisons when they have a complete octet of electrons.[4] Another form of poisoning is caused by the failure of the catalyst to desorb a product quickly. A high pressure of hydrogen often minimizes this effect by tending to increase the area of the catalyst covered with hydrogen. In the same way high pressures reduce the effect of small amounts of poisons in the reaction mixture.

The catalyst is dispersed in the solvent and is brought into close contact with the organic material and hydrogen either by shaking or stirring. Solvents which are commonly used are ethyl alcohol, acetic acid, ethyl acetate, ether, dioxane, and saturated hydrocarbons such as *n*-hexane, decalin, *cyclo*hexane and methyl*cyclo*hexane. Attempts to correlate the rate of hydrogenation with the nature of the solvent have been unsuccessful so far, although it has been shown that the product sometimes depends on the solvent. The hydrogenation of triphenylmethane in ethyl alcohol stops when two of the three aromatic nuclei have been reduced, but all three rings are saturated when methyl*cyclo*hexane is the solvent.[5]

The rate of hydrogenation and the nature of the final product can be varied by changes in the experimental conditions, although the latter is rare with olefines, as normally they reduce

[1] C. Paal and W. Hartmann, *Ber.* 1918, **51**, 711–37.
[2] P. Gaubert, R. P. Linstead and H. N. Rydon, *J. Chem. Soc.* 1937, pp. 1974–9.
[3] R. Adams and V. Voorhees, *loc. cit.* (p. 2, n. 1); cf. R. Truffault, *Bull. Soc. Chim.* 1935, [v], **2**, 244–53.
[4] E. B. Maxted and R. W. D Morris, *J. Chem. Soc.* 1937, pp. 252–6.
[5] H. Adkins, W. H. Zartman and H. Cramer, *J. Amer. Chem. Soc.* 1931, **53**, 1425–8.

under very mild conditions. Heating the reaction mixture usually increases the rate of hydrogenation, but the effect may be less than anticipated as the partial pressure of the hydrogen is reduced by a rise in the vapour pressure of the solvent. This is one way in which the solvent can affect the reaction indirectly.

An increase in pressure also increases the rate of reduction, but the effect varies greatly with the particular catalyst used and is most marked with nickel. The rate of absorption of hydrogen depends upon the surface area of the catalyst, so that the amount of solid present and its state of division are important. Complete reduction of many polyenes, such as the carotenoids, is impossible unless a relatively large amount of the catalyst is used (e.g. 0·5–2·5 g. of platinum per g. of compound). The hydrogen, the organic compound and the catalyst are brought into contact by shaking, and the rate of absorption can be affected considerably by changes in the rate of shaking.

The selectivity in action of catalysts must also be considered. Nickel is more active toward olefinic linkages than most of the oxide catalysts. The esters of unsaturated acids can be hydrogenated to unsaturated alcohols with zinc chromite at 250° C., whereas the double bond only is reduced with nickel below this temperature. The catalysts become less selective as the temperature is raised, and there is an increased probability of forming the saturated hydrocarbon.

The olefinic linkage is one of the groups reduced most readily and any of the common catalysts can be used. The simple alkenes react with hydrogen and a catalyst at room temperature and atmospheric pressure, but reduction becomes more difficult as the degree of substitution increases at the double bond.[1] A completely substituted ethylene, such as dodecahydrophenanthrene where the double bond is common to two ring systems, may be very resistant to hydrogenation.[2] Conjugation does not always retard hydrogenation, but there are many examples recorded where more drastic conditions are needed. When a double bond is conjugated to a carbonyl or carboxyl group it is

[1] W. H. Zartman and H. Adkins, *J. Amer. Chem. Soc.* 1932, **54**, 1668–74; S. V. Lebedev and M. Platonov, *J. Chem. Soc.* 1930, pp. 321–36.
[2] J. R. Durland and H. Adkins, *J. Amer. Chem. Soc.* 1938, **60**, 1501–5.

hydrogenated more slowly than an isolated double bond (e.g. the
γδ double bond in sorbic acid is easier to reduce than the αβ
bond which is in conjugation).[1] The selective reduction of a
double bond in a side chain to an aromatic ring is often possible
owing to the ease with which the olefinic linkage is hydrogenated
preferentially. The *cis* form of an ethylenic compound usually
hydrogenates more quickly than the *trans*.[2]

The breaking of a bond which can take place during hydrogenation is usually called hydrogenolysis. The conditions
generally are drastic, and normally hydrogenolysis can be
ignored for hydrogenations at room temperature and atmospheric pressure. A halogen group, however, can be replaced
under these conditions in many aliphatic compounds, and even
the relatively inert aromatic halogen compounds can be dehalo-
genated and hydrogenated to the *cyclo*paraffins with Adams'
catalyst at 3 atm. pressure and a temperature of 50–70° C. This
must be taken into account when unsaturated halides are
estimated quantitatively by hydrogenation.

A. Macro-hydrogenation

A hydrogenation apparatus is as shown in fig. 2.

The taps and ground-glass joints of the apparatus should be
examined carefully before the experiments to ensure that they
are greased properly and free from striations. Vacuum grease
only must be used.

The reaction flask *F* containing the catalyst and some of the
liquid which is used as the solvent for the olefine (20–50 ml.) is
attached to the apparatus and held firmly in position by suitable
clips (rubber bands stretched between metal holders are satisfactory). The burettes B_1 (500 ml.) and B_2 (250 ml.) are filled
with water and, with the taps T_1 and T_5 open and T_2, T_3 and T_4
closed, the apparatus is evacuated by a vacuum, or a good water,
pump connected to T_1 through a trap. The liquid in *F* is degassed
by shaking the flask whilst the air is being sucked out of the

[1] E. H. Farmer and R. A. E. Galley, *J. Chem. Soc.* 1933, pp. 687–96.
[2] C. Paal and H. Schiedewitz, *Ber.* 1927, **60**, 1221–8; 1930, **63**, 766–78;
cf. C. Weygand, A. Werner and W. Lazendorf, *J. prakt. Chem.* 1938, **151** [ii],
231–2.

apparatus. When the manometer M shows a steady state T_1 is closed and the apparatus filled with hydrogen from a cylinder which is connected to T_2 through a simple mercury valve, V, which is necessary to reduce the pressure to atmospheric. T_2 is

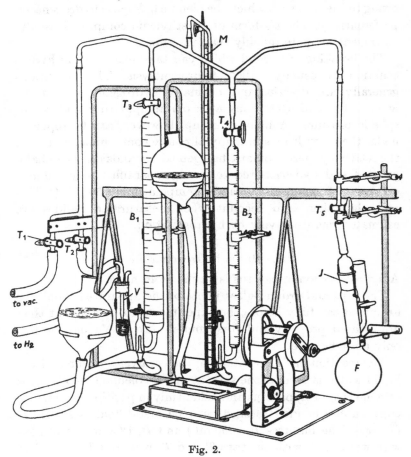

Fig. 2.

closed, T_1 is opened, the apparatus is evacuated again, T_1 and T_5 are closed and the apparatus, including the burettes, is filled with hydrogen. T_5 is opened and T_2 and T_3 are closed. The flask is shaken mechanically until no more hydrogen is absorbed from burette B_2 which corresponds to the complete reduction of the catalyst. A solution of the olefine is added to the reaction flask through the side arm of F, and great care is taken to exclude air and to wash the side arm once or twice with the pure solvent. The readings of the burettes are taken with T_3, T_4 and T_5 open and with the water at the same level in the reservoirs R_1 and R_2 and in the burettes. The tap on one of the burettes is closed (depending on the rate of absorption) and the hydrogenation is carried out, plotting a graph of the hydrogen absorbed against time. A reading of the temperature and the pressure is made and a correction is applied to the volume of gas taken up.

The olefine, solvent and catalyst may all be placed in the reaction flask at the beginning of the experiment, as the error from the absorption of hydrogen by the catalyst is usually insignificant compared with the total volume.

It is convenient sometimes to bring the reactants into contact with the catalyst by stirring. A flat-bottomed flask with a small glass-covered iron paddle inside is used as the reaction vessel and mixing is brought about by a magnet rotating under the flask.

Double bonds per mol.

$$= \frac{V \times 273 \times P \times \text{mol. wt.}}{(273 + t) \times 760 \times \text{wt. of sample} \times 22,400}.$$

Accuracy $\pm 1 \%$.

B. Micro-hydrogenation

The apparatus is as shown in fig. 3. The hydrogenation is carried out in the flask F_1 which is connected to the compensating flask F_2 by a sensitive manometer, M, filled with dibutyl phthalate. The volume of F_2 is adjusted very carefully to give accurate compensation.

The catalyst and solvent are added to F_1 and an equal volume of solvent is placed in F_2. The small tube, t, with a weighed amount of the olefine, is slipped into the side arm A of the reaction flask with the joint, J_1, turned into position to support the tube (see fig. 2b). Taps T_2, T_3, T_4, T_5 and T_6 are opened, a clip is screwed down on the rubber tubing connecting the burette B with its mercury reservoir and the apparatus is evacuated *slowly* through T_8. When a steady state is reached T_8 is closed,

F_1 and F_2 are immersed almost to the joints J_1 and J_2, either in a bath of water at room temperature or placed in a thermostat. The apparatus, including the burette, is filled with pure, dry

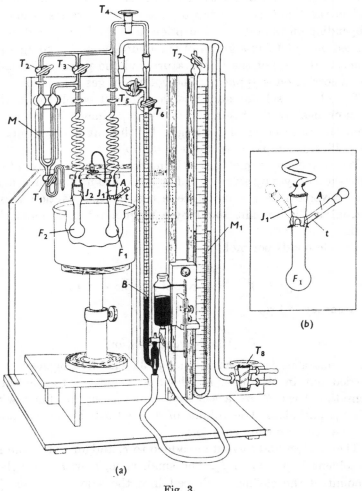

Fig. 3.

hydrogen at atmospheric pressure as measured on M, T_8 is closed, the screw-clip removed from the rubber tubing and the mercury reservoir is adjusted until a reading on the calibrated scale is possible. The reading is taken, T_5 and T_3 are closed and the flasks are shaken mechanically until the catalyst is reduced completely. The dibutyl phthalate in M is at the same level originally in both limbs, but as hydrogen is absorbed it begins to rise on the side connected to F_1. The pressure in F_1 is restored to atmospheric by constantly adjusting the level of the mercury in the burette, and as there is a great difference in the densities of the two liquids a small change in the mercury level compensates for a big rise in the liquid level in M.

When equilibrium has been established the shaking is stopped, the levels in M are adjusted carefully to the same height and the reading of the burette B is taken. The reaction flask is rotated to let the tube, t, drop into the flask. This should be done quickly and with the minimum of handling to reduce as far as possible the transfer of heat from the hand. A slight change in the level of M is inevitable and should be ignored. The flasks are shaken again and the reduction is carried out in the same way as for the catalyst. The volume of hydrogen absorbed is corrected for temperature and pressure (see p. 11).

Accuracy $\pm 2\%$.

BROMINE NUMBER

$$>\!C\!=\!C\!<\; + Br_2 = >\!\underset{Br\;\;Br}{C\!-\!C}\!<$$

A. Direct Titration

The bromine number is expressed as the weight of bromine absorbed by 100 g. of the substance (cf. iodine number, p. 20).

A solution of bromine in a solvent such as carbon tetrachloride or acetic acid is a vigorous reagent and substitution as well as addition usually takes place, although the degree depends upon the type of structure. Highly branched olefines are generally substituted very heavily. A correction can be made for the amount of substitution, as in this reaction an equivalent of hydrogen bromide is formed and can be found by titration with alkali.[1] These corrected values are sometimes negative

---

[1] P. C. McIlhiney, *J. Amer. Chem. Soc.* 1894, **16**, 275–8; 1899, **21**, 1084–9; 1902, **24**, 1109–14; H. Schweitzer and E. Lungwitz, *J. Soc. Chem. Ind.* 1895, **14**, 130–3.

14   OLEFINES

either from incomplete addition, or from the liberation of more than one equivalent of acid (see p. 16), and this method therefore can only be used as a rapid means of comparison and not as an absolute measure of unsaturation (see below). The reagents used in this estimation must be neutral and the solvents must be dried carefully before they are used. This latter condition is difficult with carbon tetrachloride as the solvent.

1. *Titration with potassium bromide-potassium bromate solution*

Substitution can be limited and frequently reduced to negligible amounts by using an acidified bromide-bromate solution.[1] The rate of bromine formation depends on the strength of the acid, so that by making the reaction mixture just acid and by shaking vigorously the bromine concentration is always kept low enough to avoid substitution. This method has been found to give good results with straight-chain olefines but is unreliable for cyclic structures such as indene, furane and phenylbutadiene.[2] Many modifications to the original method have been suggested,[3] and the estimation described below[4] gives accurate values for a wide range of unsaturated compounds. The bromine numbers of the straight-chain olefines sometimes are slightly lower (by about 1–2 %) than the theoretical when found by this method, whilst those of the branched chain higher olefines, the cyclic olefines, such as *cyclo*hexene, indene and 1:4-dihydronaphthalene, are almost equal to the theoretical value. Nonconjugated diolefines, 4-vinyl*cyclo*hexene and *d*-limonene also give good results. The aromatic and paraffin hydrocarbons in the gasoline and kerosene boiling ranges do not react with bromine under the conditions described, but anthracene in the higher boiling ranges, and probably its derivatives, undergo substitution. Stilbene and related olefines only absorb bromine

[1] A. W. Francis, *J. Ind. Eng. Chem.* 1926, **18**, 821–2.

[2] F. Cortese, *Rec. trav. chim.* 1929, **48**, 564–7.

[3] F. S. Bacon, *J. Ind. Eng. Chem.* 1928, **20**, 970–1; S. P. Mulliken and R. L. Wakeman, *Ind. Eng. Chem.* (Anal. ed.), 1935, **7**, 59; C. L. Thomas, H. S. Bloch and J. Hoekstra, *ibid.* 1938, **10**, 153–6; J. B. Lewis and R. B. Bradstreet, *ibid.* 1940, **12**, 387–90; S. J. Green, *J. Inst. Petroleum*, 1941, **27**, 66–71; H. Grosse-Oetringhaus, *Brennstoff-Chem.* 1938, **19**, 417–27; K. Uhrig and H. Levin, *Ind. Eng. Chem.* (Anal. ed.), 1941, **13**, 90–2.

[4] H. L. Johnson and R. A. Clark, *Anal. Chem.* 1947, **19**, 869–72.

slowly and give low results, and conjugated dienes give a mixture of addition products so that the values are intermediate between one and two double bonds.

The olefine (1–2 g. for bromine numbers between 50 and 200, and 5 g. for bromine numbers of 20 and below) is weighed into a graduated flask (50 ml.) containing carbon tetrachloride (25 ml.), and a solution of known concentration is made up (to 50 ml.) with the same solvent. A volume (5 ml.) is measured into a glass-stoppered bottle (16 oz.) and dissolved in glacial acetic acid (50 ml., B.P.). The bromide-bromate reagent (see below) is added (1–2 drops each second) from a burette until a distinct yellow colour persists during 5 sec. and then an excess of the reagent (1 ml.) is added. This titration should be made at $25 \pm 5°$ C. The contents of the flask should be kept mixed by a swirling motion, and they must not be exposed to direct sunlight. The stopper is placed in the bottle and the mixture is shaken during 40 ± 5 sec. An aqueous solution of potassium iodide (5 ml., 15 %) and distilled water (100 ml.) are added immediately and the bottle is shaken vigorously during 1 min. and titrated straight away with a standard solution of sodium thiosulphate (0·1 N). Starch (1 ml., 0·5 %) is added near the end-point. The possible loss of bromine vapour which can occur when the stopper is removed is prevented by placing the solution of potassium iodide in the lip of the bottle and taking out the stopper gradually.

Bromine number

$$= \frac{(\text{ml. of KBr-KBrO}_3 \times \text{N} - \text{ml. of Na}_2\text{S}_2\text{O}_3 \times \text{N}) \times 7 \cdot 992}{\text{weight of the olefine (in g.)}}.$$

Accuracy $\pm 1·5 \%$.

PREPARATION: *Potassium bromide-potassium bromate solution.* Potassium bromide (49·6 g.) and potassium bromate (13·9 g.) are dissolved in water, mixed and the solution made up (to 1 l.).

The solution is standardized by measuring a volume (5 ml.) into a glass-stoppered bottle (16 oz.), dissolving in glacial acetic acid (50 ml.) and leaving the mixture to stand for 5 min. An aqueous solution of potassium iodide (5 ml., 15 %) and distilled water (100 ml.) are added, and the liberated iodine is titrated with a standard solution of sodium thiosulphate (0·1 N) as above.

2. *Titration with bromine in carbon tetrachloride*

A solution of the olefine is made up as described under sub-head 1 (p. 15). A volume (10 ml.) is measured into a glass-stoppered bottle (16 oz.), and a solution of bromine in dry carbon tetra-chloride (25 ml., 0·2 N) is added. The stopper is placed in the bottle and the mixture is kept at 20° C. for half an hour. The bottle is cooled in an ice-bath, and an aqueous solution of potassium iodide (10 ml., 15 %) and distilled water (100 ml.) are added as described on p. 15. The liberated iodine (A) is titrated as above (p. 15).

The contents of the bottle are transferred to a separating funnel, the aqueous layer is separated and titrated with a standard solution of potassium hydroxide using methyl orange as the indicator (B).

The bromine equivalent to addition is $A - 2B$.

B. Electrometric Titration

The addition of bromine is catalysed by silver, aluminium, nickel, uranium, zinc and mercury salts, and when mercuric chloride is present direct titration of olefines is possible with standard solutions of either bromine in carbon tetrachloride[1] or potassium bromide-potassium bromate.[2] Substitution and oxidation reactions are prevented almost entirely as the reagent is never in excess, and in the estimation with potassium bromide-potassium bromate the risk of substitution is reduced still further by cooling the reaction mixture. The end-point is found electrometrically, which is of considerable help when the solutions are highly coloured.[3]

[1] B. Braae, *Anal. Chem.* 1949, **21**, 1461–5; cf. H. J. Lucas and D. Pressman, *Ind. Eng. Chem.* (Anal. ed.), 1938, **10**, 140–3.

[2] H. B. DuBois and D. A. Skoog, *ibid.* 1948, **20**, 624–7.

[3] Cf. p. 51 and C. W. Foulk and A. T. Bawden, *J. Amer. Chem. Soc.* 1926, **48**, 2045–50.

1. *Titration with bromine in carbon tetrachloride*

The olefine (1–3 m.equiv. which should be equivalent to not more than 35 ml. of the bromine solution) is weighed into a beaker (100 ml.) and the methyl alcoholic reagent solution is added (50 ml., see below). When the olefine is not easily soluble in methyl alcohol it can be taken up in chloroform, or in carbon tetrachloride (a few ml.). A mechanical stirrer is placed in the solution together with two electrodes (0·5 mm. platinum wires which must be covered completely) connected to an amplifier

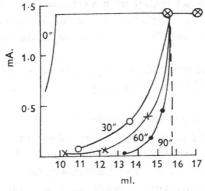

Fig. 4.

and to a suitable potential. The ammeter in the circuit is adjusted to zero and then the olefine is titrated slowly (0·5 ml. lots) with a solution of bromine in dry carbon tetrachloride (8 g. diluted to 1 l., 0·1 N, see below).

A stopwatch is started at the beginning of the titration and the current between the electrodes is read on the ammeter after 30, 60, 90 and 120 sec. The titration is continued until the current is constant, which is taken as the end-point.

The addition of a small amount of bromine to the olefine allows a current to pass between the electrodes which gradually falls to its original level as reaction takes place. A graph of the reagent added against the ammeter readings gives the end-point as the point of intersection of the current curves after 0, 30, 60, 90 and 120 sec. and the volume of bromine required corresponds to the constant value of the current (see fig. 4).

A blank titration is done and should be negligible.

The titration can be done as described in method 2 below.

Accuracy approximately ± 2 %.

PREPARATIONS: *The methyl alcohol solution.* A solution of hydrogen bromide (15 ml., sp.gr. = 1·38) is added to concentrated hydrochloric acid (170 ml., sp.gr. = 1·19) and mercuric chloride (20 g.) and diluted (to 1 l.) with pure methyl alcohol.

The solution corresponds to 0·1N HBr and 2N HCl.

Standardization of bromine in carbon tetrachloride. The solution is standardized by adding potassium iodide, and the iodine which is liberated is titrated with a standard solution of sodium thiosulphate (0·1N). The solution should be stored in the dark and, if needed for use during several days, its strength should be found each day.

The solution is unstable if sulphur is present in the carbon tetrachloride used for making it up.

2. *Titration with potassium bromide-potassium bromate solution*

The olefine (0·2 g. for bromine numbers greater than 200, 0·5 g. for bromine numbers between 20 and 200 and 2·0 g. for bromine numbers between 1 and 20) is weighed, or measured by a pipette, into the titration vessel containing the solvent (110 ml., see below) cooled to 5° C. If the sample is viscous and difficult to transfer, it can be taken up in carbon tetrachloride (a few ml.). The solution is stirred mechanically, cooled to between 0° and 5° C. and the 'magic eye' of an electronic indicating tube adjusted until it is nearly closed. The potassium bromide-potassium bromate reagent (see below) is added at first drop by drop, and if the 'magic eye' does not flutter it is run in at a faster rate. The titration is continued by small additions until the 'magic eye' remains open for 30–45 sec. If more than 15 ml. of the reagent is used the titration is repeated using a smaller sample.

$$\text{Bromine number} = \frac{\text{ml. of reagent} \times \text{N} \times 8}{\text{wt. of sample}}.$$

Accuracy approximately ± 2 %.

PREPARATIONS: *Potassium bromide-potassium bromate* (0·626N). Potassium bromide (270 g.) and potassium bromate

(17·42 g.) are dissolved in water, mixed, and the solution made up (to 1 l.).

The solution is standardized as described on p. 15.

The solvent for the olefine. The solvent is made by mixing glacial acetic acid (80 vol., B.P.), methyl alcohol (7 vol.) carbon tetrachloride (15 vol., B.P.), sulphuric acid (2 vol., 6 N) and an alcoholic solution of mercuric chloride (2 vol., 10 g. per 100 ml. of ethyl alcohol).

C. Bromine Absorption

The bromine value of unsaturated oils can be found by spreading a thin film of the oil on a microscope slide and exposing to bromine vapour.[1] The weight of the film is found before and after bromination. The method is rapid, the values agree well with those determined by other methods, the appearance of the film is frequently characteristic of a particular oil and substitution can be prevented by doing the reaction in the dark. Most unoxidized oils give results which agree with the iodine number as found by Wijs' method, but a higher absorption is found for oils containing oxidized acids (boiled linseed oil). Castor oil, containing ricinoleic acid, shows a lower unsaturation value than that found with iodine monochloride. The three conjugated double bonds in α-elaeostearic acid

$$(C_4H_9{-}(CH{=}CH)_3{-}(CH_2)_7 . COOH)$$

are saturated, and the method gives a direct measure of the unsaturation in tung oil which contains 60 % or more of this acid (see p. 21).

A microscope slide is weighed before and after a drop of the oil (0·02–0·03 g.) is spread on its surface as a thin film (about 0·2 mm. thick). The slide, together with a boat containing bromine (1–2 drops), is placed in a wide tube which is closed at each end by a waxed cork. The slide is taken out after 20–30 min., and the excess bromine is removed either by heating at 50–60° C. or by passing a current of warm air or nitrogen over

[1] H. Thoms, *Analyst*, 1928, **53**, 69–77; P. Becker, *Z. angew. chem.* 1923, **36**, 539; E. Rossmann, *Ber.* 1932, **65**, 1847–51; J. Böeseken and P. Pols, Jr., *Rec. trav. chim.* 1935, **54**, 162–6; Th. Sabalitschka and K. R. Dietrich, *Pharm. Ztg.* 1924, **69**, 425–6; O. Helner, *Analyst*, 1895, **20**, 49–52.

the surface. The slide is reweighed and the bromine number is found from the increase in weight.

More of the oil is needed when the bromine number is low, and it is advisable then to take a larger plate and use a similar one as a counterpoise for weighing.

IODINE NUMBER

$$\text{C=C} + I_2 = \text{C—C}$$

The iodine number is expressed as the weight of iodine absorbed by 100 g. of the substance. It has been used for a very long time as a measure of the degree of unsaturation and is probably the best-known constant in the analysis of fats and oils. High values are characteristic of the liquid vegetable fats, and the number falls from 206, for perilla oil, to between 5 and 25 for the coconut oils. The number may vary with the source of supply, as conditions under which the oilseed is grown affects the composition.

Many methods have been described by which the value can be found, and of these the methods of Wijs, Hanus and Rosenmund-Kuhnhenn are used most frequently. The older and more tedious estimation of Hübl is not used very often. The main reaction in all the methods is the addition of halogen, but small amounts of the substitution products may be formed. Olein gives di-iodostearin by the addition of six atoms of iodine, but palmitin under similar conditions only forms a small amount of iodopalmitin by substitution. The methods therefore fail to differentiate between addition and substitution, but by controlling the experimental conditions very carefully reproducible results are obtained although the values depend on the method used. The iodine number as found using Hanus' solution is some 2–4 % lower than by the Wijs or Rosenmund-Kuhnhenn methods for all oils having a number greater than 100.[1]

[1] L. M. Tolman and L. S. Munson, *J. Amer. Chem. Soc.* 1903, **25**, 244–51; F. R. Earle and R. T. Milner, *Oil and Soap*, 1939, **16**, 69–71.

A. Wijs' Method

An excess of an acetic acid solution of iodine monochloride is added to a weighed amount of the olefine dissolved in chloroform or carbon tetrachloride, and when no further absorption takes place the excess is determined.[1] The procedure has been adopted as an official method of estimation of oils, fats and petroleum products.[2] Pure, non-conjugated, fatty acids give quantitative results, but compounds having conjugated double bonds or acetylenic linkages only take up the full amount of halogen with difficulty and variable, unreliable values are obtained.[3] Tung oil gives anomalous results for this reason. The observed iodine number is about two-thirds of the theoretical value owing to the α-elaeostearic acid which is combined with the oil. An empirical figure which does not correspond to full absorption can be found by using three times the normal amount of the reagent, and leaving the reaction mixture to stand for 3 hr.[4] The method is now of little use, as a value very close to the theoretical for the three conjugated double bonds can be found by the absorption of bromine (see pp. 19 and 25). Other methods have been described in which absorption is limited to two of the double bonds.[5] Low iodine numbers are found for the vinyl esters of the fatty acids from caproic to stearic acids, but good results are possible if the normal method is modified so that the reaction mixture of the olefine and reagent either is left to stand for 24 hr. instead of 30 min. or a 20% excess of the reagent is used and left for 1 hr.[6] The method has been used for the estimation of unsaturation in natural products. The results are

---

[1] J. J. A. Wijs, *Ber.* 1898, **31**, 750–2; *Z. angew. Chem.* 1898, **11**, 291–7; *J. Soc. Chem. Ind.* 1898, **17**, 698; *Chem. Rev. Fett. Harz. Ind.* 1899, **6**, 29–31; *Analyst*, 1929, **54**, 12–14.

[2] Assoc. Official Agr. Chem., *Official and Tentative Methods of Analysis*, 6th ed. 1945, p. 495; *Standard Methods for Testing Petroleum and its Products*, 7th ed. 1946, pp. 191–3; A.C.S. Committee on the Analysis of Commercial fats and oils, *J. Ind. Eng. Chem.* 1926, **18**, 1354–5.

[3] J. van Loon, *Chem. Umschau Fette, Oele, Wachse u. Harze*, 1930, **37**, 85–7, 257–62.

[4] A. C. Chapman, *Analyst*, 1912, **37**, 543–53.

[5] E. R. Bolton and K. A. Williams, *Analyst*, 1930, **55**, 360–5; *Oil and Soap*, 1938, **15**, 315–6; J. D. von Mikusch, *ibid.* 1938, **15**, 186–8.

[6] D. Swern and E. F. Jordan, Jr., *J. Amer. Chem. Soc.* 1948, **70**, 2334–9.

22   OLEFINES

erratic with the sterols and unsaponifiable matter,[1] but reasonably accurate values can be found for the carotenoids.[2] The unsaturation of natural and synthetic rubbers can be determined by an elaborate modification of Wijs' method.[3] When an excess of iodine monochloride is added to the polymer both addition and substitution take place with a breakdown of the molecule to give an acid. Corrected iodine values are calculated and give a measure of the unsaturation.

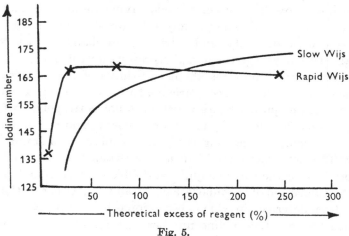

Fig. 5.

The standard method of estimation and a recent modification are described. In the former, a reaction time of 30–60 min. is usually necessary to ensure complete addition, and a large excess of the reagent must be used for the saturation of all the double bonds. Mercuric acetate in glacial acetic acid catalyses the addition, and saturation of the double bonds is possible within 5 min., using a much smaller excess of the reagent.[4] The iodine numbers found by the latter procedure are lower than by the standard method but agree with results using Hanus' solution (see fig. 5).

[1] E. R. Bolton and K. A. Williams, *Analyst*, 1930, **55**, 5–11.
[2] R. Pummerer, L. Rebmann and W. Reindel, *Ber.* 1929, **62**, 1411–18.
[3] T. S. Lee, I. M. Kolthoff and M. A. Mairs, *J. Polymer Sci.* 1948, **3**, 66–84.
[4] H. D. Hoffman and C. E. Green, *Oil and Soap*, 1939, **16**, 236–8; F. A. Norris and R. J. Buswell, *Ind. Eng. Chem.* (Anal. ed.), 1943, **15**, 258–9; D. J. Hiscox, *ibid.* 1948, **20**, 679–80.

Hanus' solution is easier to prepare than Wijs' and it is stable for one to two years, whereas it has been recommended that Wijs' solution should not be used after 30 days.[1] This deterioration of iodine monochloride is said to be much less than was thought originally, as it can be kept without change just as long as Hanus' solution.[2]

1. Standard method of analysis

The olefine (0·1–0·5 g., depending on the iodine number) is weighed into a clean, dry, glass-stoppered bottle (16 oz.) and dissolved in chloroform or carbon tetrachloride (15–20 ml.). The reagent (25 ml., 50–60 % excess, i.e. 100–150 % of the quantity absorbed) is added, the stopper is moistened with an aqueous solution of potassium iodide (15 %) to prevent the loss of halogen, and the mixture is left to stand in the dark at a uniform temperature for 30 min. An aqueous solution of potassium iodide (20 ml., 15 %) and water (100 ml., freshly boiled and cooled) are added, and the liberated iodine is titrated with a standard solution of sodium thiosulphate (0·1 N). Starch (1 ml., 0·5 %) is added near the end-point.

Two blank determinations should be done in parallel. The bottles should all be kept at the same temperature, as a small difference can affect the titration value considerably.

Iodine is precipitated in the estimation of tung oil.

2. With mercuric acetate added

The olefine (0·1–0·5 g., depending on the iodine number) is weighed into a clean, dry, glass-stoppered bottle (16 oz.) and dissolved in chloroform or carbon tetrachloride (10 ml.). The reagent (10 ml.) and a solution of mercuric acetate in glacial acetic acid are added (5 ml., 2·5 %) and the mixture left to stand for 5 min. (6 min. for oil of flaxseed). An aqueous solution of potassium iodide (5 ml., 15 %) and water (25 ml.) are added, and the iodine is titrated as described under sub-head 1.

[1] Assoc. Official and Agr. Chem. *Official and Tentative Methods of Analysis*, 6th ed. 1945, p. 495; A.C.S. Committee on the Analysis of Commercial fats and oils, *loc. cit.* p. 21, n. 2.
[2] T. P. Hilditch, *Industrial Chemistry of Fats and Waxes*, 1941, p. 48; F. A. Norris and R. J. Buswell, *Ind. Eng. Chem.* (Anal. ed.), 1944, **16**, 417.

PREPARATION: *Wijs' solution.* Resublimed iodine (13·0 g.) is dissolved in acetic acid (1 l., 99·5 %, see below). It may be necessary to warm the mixture on a water-bath. The solution is cooled and divided into two parts (900 and 100 ml.). Pure, dry chlorine is passed into the larger volume until the colour has changed from deep brown to a dark orange. The iodine in acetic acid is added until the solution on mixing becomes faintly brown. The halogen content of the reagent and of the original iodine solution is determined by titration with a standard solution of sodium thiosulphate (0·1 N); the ratio of the values should be slightly less than two. An excess of iodine (about 1·5 %) should be present and no excess of chlorine.

The reagent is stored in a dark amber, glass-stoppered bottle which is sealed with paraffin wax until it is used.

Iodine trichloride should not be used for the preparation of the reagent as it is very unstable.

The acetic acid used in preparation should be pure (at least 99·5 %) and should show no reduction with potassium dichromate and sulphuric acid (cf. p. 31).

B. Hanus' Method

Hanus' method using iodine monobromide is very similar to Wijs', and frequently it is preferred to the latter, as the reagent is easier to prepare than the monochloride and there is no doubt about its stability (see p. 23).[1] The reaction can be catalysed by mercuric acetate in acetic acid and, except for castor oil which contains ricinoleic acid, the results are identical with those found by the standard procedure (cf. p. 19). With castor oil, however, more of the reagent is used, as mercuric acetate also catalyses a slow reaction at the hydroxyl group of ricinoleic acid. O-propionyl ricinoleates give the same value by either method.[2] Fatty acids with acetylenic or conjugated double bonds are saturated only with difficulty, and the results are very unreliable from the catalysed reaction as the iodine number varies with the excess of the reagent used. There is no

[1] J. Hanus, *Z. Untersuch. Nahr.- u. Genussm.* 1901, **4**, 913–20.
[2] P. S. Skell and S. B. Radlove, *Ind. Eng. Chem.* (Anal. ed.), 1946, **18**, 67–8; F. A. Norris and R. J. Buswell, *ibid.* 1943, **15**, 258–9.

addition with α-methylene butyrolactone and this may be general for all αβ-lactones.[1]

A micro-method of estimation using a single drop of oil has been described.[2]

1. *Standard method of analysis*

The olefine (0·1–0·5 g., depending on the iodine number) is weighed into a clean, dry, glass-stoppered bottle (16 oz.) and dissolved in chloroform or carbon tetrachloride (10 ml.). The reagent (25 ml., at least 60 % excess) is added and the mixture is left to stand for 30 min. exactly. An aqueous solution of potassium iodide (10 ml., 15 %) and water (100 ml., freshly boiled and cooled) are added and the liberated iodine is titrated with a standard solution of sodium thiosulphate (0·1 N). Starch (1–2 ml., 1 %) is added near the end-point and the mixture is shaken vigorously after each addition of sodium thiosulphate.

Two blank determinations should be done in parallel.

2. *With mercuric acetate added*

The olefine (0·1–0·5 g., depending on the iodine number) is weighed into a clean, dry, glass-stoppered bottle (16 oz.) and dissolved in chloroform or carbon tetrachloride (10 ml.). The reagent (25 ml.) and a solution of mercuric acetate (10 ml., 2·5 %) in glacial acetic acid are added and, after shaking, the mixture is left to stand for 3–5 min. An aqueous solution of potassium iodide (5 ml., 15 %) and water (25 ml.) are added and the iodine is titrated as described under sub-head 1.

Two blank determinations should be done in parallel with the estimation. These values are consistently 2–3 % lower than those found in method 1 above.

PREPARATION: *Hanus' solution*. Resublimed iodine (13·625 g.) is dissolved in acetic acid (825 ml., 99·5 %, see below). It may be necessary to warm the mixture on a water-bath. The solution is cooled and titrated (25 ml.) with a standard solution of sodium thiosulphate (0·1 N), using starch as an indicator near the end-point. A solution of bromine (3 ml.) in acetic acid (200 ml.,

[1] C. J. Cavallito and T. H. Haskell, *J. Amer. Chem. Soc.* 1946, **68**, 2332–4.
[2] A. H. Gill and H. S. Simms, *J. Ind. Eng. Chem.* 1921, **13**, 547–52.

99·5 %) is standardized by adding potassium iodide (10 ml., 15 %) and titrating the liberated iodine with the standard solution of sodium thiosulphate (0·1 N). The volume of the bromine solution which must be added to the iodine solution (800 ml.) to double the halogen content is found from

$$\frac{\text{sodium thiosulphate equiv. to 1 ml. of iodine solution}}{\text{sodium thiosulphate equiv. to 1 ml. of bromine solution}} \times 800.$$

This volume of bromine in acetic acid is added to the iodine solution (800 ml.).

The reagent is stored in a dark brown, glass-stoppered bottle which is sealed with paraffin wax until it is used.

The acetic acid should be pure (at least 99·5 %) and should show no reduction with potassium dichromate and sulphuric acid (cf. p. 31).

C. Rosenmund-Kuhnhenn's Method

Pyridine bromine sulphate is a milder reagent than bromine and possibly than iodine monochloride, and sometimes it is to be preferred to other reagents which give variable results.[1] Wijs' method gives erratic values with some of the unsaturated sterols, such as cholesterol, cholestenone and ergosterol, and with the crude unsaponifiable matter of oils, whereas reliable values are found by the method of Rosenmund and Kuhnhenn.[2] The reagent is very stable, it is prepared quickly, it is easy to use, and it has found wide application in the micro-estimation of biological materials.

A solution of pyridine bromine sulphate (10 ml., see below) is added to the olefine (0·02 g.) dissolved in chloroform (5 ml.), and after shaking, the mixture is left to stand for 5 min. at room temperature. An aqueous solution of potassium iodide (3·5 ml., 10 %) is added and the iodine is titrated as described under subhead 1 (see p. 25).

[1] K. W. Rosenmund and W. Kuhnhenn, Z. Untersuch. Nahr.- u. Genussm. 1923, 46, 154–9.
[2] H. Dam, Biochem. Z. 1924, 152, 101–10; I. Smedley-MacLean, Biochem. J. 1928, 22, 23–6; A. M. Copping, ibid. 1928, 22, 1142–4; E. R. Bolton and K. A. Williams, Analyst, 1930, 55, 5–11.

PREPARATION: *The pyridine bromine sulphate solution.*
A solution of bromine (8 g.) in glacial acetic acid (20 ml.) is added
to pure, dry pyridine (8 g.) in a mixture of concentrated sulphuric
acid (10 g.) and glacial acetic acid (20 ml.). The reagent is diluted
(to 1 l.) with glacial acetic acid.

THIOCYANOGEN NUMBER

$$>\!\!C\!\!=\!\!C\!\!< + (SCN)_2 = \underset{\underset{NCS}{\,|\,}}{>}\!\!C\!\!-\!\!\underset{\underset{SCN}{\,|\,}}{C}\!\!<$$

The thiocyanogen number is expressed as the weight of thiocyanogen absorbed by 100 g. of the substance. The reaction of
thiocyanogen with unsaturated fatty acids and glycerides has
been developed as a measure of unsaturation, and the technique
has found wide use in the analysis of fats and oils.[1] An excess of
a standardized solution of thiocyanogen is added to the unsaturated compound and the amount of unreacted reagent is
found iodometrically.

The addition reaction has been shown to be quantitative for
monoethylenic fatty acids, and the values obtained agree with
the iodine number.[2] With linoleic and linolenic acids, however,
the value is less than the theoretical and corresponds to one of
the two double bonds in linoleic acid and two of the three double
bonds in linolenic acid (see below). Since iodine adds to all the
double bonds in these acids determinations by iodine and
thiocyanogen allow the composition of a mixture of oleic,
linoleic and linolenic acids to be calculated.[3] Acetylenic fatty
acids such as stearolic and behenolic acids do not seem to add

[1] H. P. Kaufmann, *Ber. pharm. Ges.* 1923, **33**, 139–48; *Arch. Pharm.* 1925,
pp. 1–47; *Z. Untersuch. Lebensm.* 1926, **51**, 15–27; *Seifens.-Ztg.* 1928, **55**,
297–300; H. P. Kaufmann and M. Keller, *Z. angew. Chem.* 1929, **42**, 20–3 and
73–6; H. P. Kaufmann and C. Lautenberg, *Ber.* 1929, **62**, 392–401; H. N.
Sher and R. H. Coysh, *Analyst*, 1939, **64**, 814–16; T. P. Hilditch and K. S.
Murti, *ibid.* 1940, **65**, 437–46; W. S. Martin and R. C. Stillman, *Oil and Soap*,
1933, **10**, 29–31; M. G. Lambou and F. G. Dollear, *ibid.* 1945, **22**, 226–32;
ibid. 1946, **23**, 97–101; Am. Oil Chemists' Soc., *Official and Tentative Methods*,
1938, p. 446; Assoc. Official Agr. Chem., *Official and Tentative Methods of
Analysis*, 6th ed., 1946, pp. 495–6.
[2] H. P. Kaufmann, *Z. Untersuch. Lebensm.* 1926, **51**, 15–27.
[3] F. T. Walker, *J. Oil Colour Chem. Assoc.* 1945, **28**, 119–34; M. A. P.
Campos, *Rev. quím. farm.* (Rio de Janeiro), 1946, **11**, no. 12, 25–7.

thiocyanogen under the conditions used and, because of this specificity, estimation by this method has gained considerable importance in fat and oil chemistry.

Variable results have been reported by many investigators apparently using the same procedure and as far as possible identical experimental conditions.[1] The procedure has been examined critically and it is not as simple as it seemed originally, as many sources of error have been shown to be present in previous methods.[2] It has been shown that when two double bonds are present the thiocyanogen value is slightly more than one, and with three double bonds the value is slightly less than two. This has not invalidated the method, but the stoichiometric relationship as given by the above equation[3] has been replaced by a set of empirical constants.[4] A comprehensive study of the experimental conditions and of the method of preparation of the reagent has been made.[5] The possible errors are the incomplete solubility in the reaction mixture of certain fats, the formation of resin-like substances which tend to occlude iodine, the addition of insufficient potassium iodide to the reaction mixture, the presence of unsuspected traces of moisture in the reaction mixture, the poor stability and variable rate of deterioration of the reagent on storage, and the source and preparation of lead thiocyanate. The modifications which have been proposed to prevent these errors are the careful preparation of thiocyanogen in a mixture of acetic acid and carbon tetrachloride, a new method of preparation of lead thiocyanate, the addition of at least twice the theoretical amount of potassium iodide to the reaction mixture, the careful drying of all glassware and solvents and keeping moisture out of the reaction mixture during the estimation. The instability of the thiocyanogen

---

[1] J. van Loon, Z. Untersuch. Lebensm. 1930, 60, 320–7; H. van der Veen and J. van Loon, Chem. Umschau Fette, Oele, Wachse u. Harze, 1932, 39, 56–9; P. J. Gay, J. Soc. Chem. Ind. 1932, 51, 126–9T; J. P. Kass, H. G. Loeb, F. A. Norris and G. O. Burr, Oil and Soap, 1940, 17, 118–19; J. P. Kass, W. O. Lundberg and G. O. Burr, ibid. 1940, 17, 50–3; R. W. Riemenschneider and D. H. Wheeler, ibid. 1939, 16, 207–9 and 219–21.
[2] M. G. Lambou and F. G. Dollear, loc. cit. p. 27, n. 1.
[3] H. P. Kaufmann, Analyst, 1926, 51, 157–8.
[4] V. C. Mehlenbacher, Chem. and Eng. News, 1944, 22, 606–8.
[5] M. G. Lambou and F. G. Dollear, loc. cit. p. 27, n. 1.

solution in acetic acid is due to reducing substances or to traces
of moisture in some, or at least one, of the components, and also
to the necessity of storing the reagent above the freezing-point
of the solvent. Reducing substances and water can be removed
before the preparation of the reagent, and by using a mixed
solvent of acetic acid and carbon tetrachloride the solution can
be stored conveniently at 5° C. The reagent as prepared and
stored under these conditions deteriorates slowly (see fig. 6). The

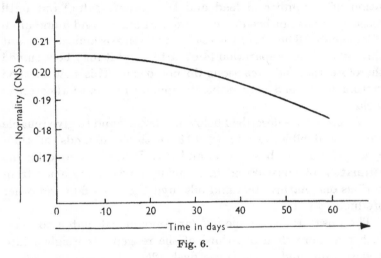

Fig. 6.

majority of fats are soluble in the mixed solvent, and it is better
to incorporate carbon tetrachloride in the reagent rather than
add it to the reaction mixture as the latter procedure reduces
the normality of the reagent. The use of a mixed solvent is not
a new suggestion but has been tried frequently in this estimation.[1] When no carbon tetrachloride is used resinous products
may be formed when water is added, and only by long and
vigorous shaking is it possible to estimate the whole of the iodine
liberated. With compounds of low thiocyanogen number larger
amounts of the substance are weighed out, and without carbon

[1] H. P. Kaufmann, 'Studien auf dem Fettgebiet', *Verlag Chemie*, Berlin,
1935; H. P. Kaufmann and L. Hartweg, *Fette und Seifen*, 1938, **45**, 346–9;
Assoc. Official Agr. Chem., *Official and Tentative Methods of Analysis*, 6th ed.
pp. 495–6; N. L. Matthews, W. R. Brode and J. B. Brown, *Oil and Soap*, 1941,
18, 182–7.

tetrachloride a thick scum often forms. The amount of potassium iodide now recommended is at least twice the theoretical amount which is much more than was used previously.[1] Powdered potassium iodide is added to the reaction mixture and then it is dissolved in water. The advantages of this method are not accepted universally, and the use of an aqueous solution of potassium iodide is often preferred.

Lead thiocyanate has been prepared by the double decomposition of a number of lead and thiocyanate salts,[2] but until recently it has not been made from lead nitrate and ammonium thiocyanate.[3] The pH of the reaction mixture is almost constant throughout the preparation (4·07–4·13), and it may be assumed therefore that side reactions do not occur. This new method gives a pure, relatively stable compound in almost theoretical yield.

The procedure described below has been found to give reliable and reproducible results (± 0.13) with compounds of thiocyanogen number between 14 and 167. It has been used for the estimation of unsaturated fatty acids, mixed fatty acids from various oils, natural fats and oils, hydrogenated fats and other products.

The unsaturated compound (an amount calculated to react with not more than one-third of the reagent) is weighed into a glass-stoppered Erlenmeyer flask (250 ml.). Excess (200%, 0·2N, see below) of the reagent is added and the mixture is left to stand in the dark for 24 hr. at room temperature (19° C.). Three blanks should be made up at the same time under identical conditions for each short series of estimations. Finely powdered potassium iodide (twice the theoretical amount, see below) is added and mixed by swirling (for 2 min. for the estimation and 3 min. for the blanks). Distilled water (20 ml.) is added and mixing is continued until all the solid has dissolved. The

[1] Cf. V. C. Mehlenbacher et al. 'Report of the Committee on Analysis of Fats and Oils', *Ind. Eng. Chem.* (Anal. ed.) 1945, **17**, 336–40.

[2] Amer. Oil Chemists' Soc. *Official and Tentative Methods*, 1938, p. 446; Assoc. Official Agr. Chem. *Official and Tentative Methods for Analysis*, 6th ed. pp. 495–6; H. S. Booth, *Inorganic Syn.* **1**, 85–6; Z. Karaoglanov and B. Sagortschev, *Z. anorg. allgem. Chem.* 1931, **202**, 62–72; H. P. Kaufmann and H. Grosse-Oetringhaus, *Oel, Kohle, Erdoel, Teer*, 1938, **14**, 199–201.

[3] M. G. Lambou and F. G. Dollear, *Oil and Soap*, 1946, **23**, 97–101.

liberated iodine is titrated with a standard solution of sodium thiosulphate (0·1N) adding starch (1 ml., 1 %) near the end-point.

PREPARATIONS: *Lead thiocyanate.*[1] Aqueous solutions of lead nitrate (500 ml., M; i.e. 331 g. per l.) and ammonium thiocyanate (136 ml., 10M; i.e. 760 g. per l.) are filtered and cooled to 5° C. The solutions are mixed by adding the lead nitrate gradually (in 50 ml. lots) to the ammonium thiocyanate with moderate stirring which is continued during 30 min. after the final addition. The copious white precipitate which separates is washed free from nitrates by decantation with distilled water cooled to 5° C. The precipitate is transferred to a Büchner funnel and pressed and sucked as dry as possible. The solid is dried finally by standing during 6–8 days in a vacuum desiccator charged with phosphorus pentoxide. The lead thiocyanate must be protected from direct sunlight throughout the preparation and during storage. If these precautions are taken the solid is sufficiently stable to store over phosphorus pentoxide for at least 18 months. Yield: 156 g. (97 % on lead nitrate).

Carbon tetrachloride. Carbon tetrachloride is shaken in a separating funnel with successive portions of concentrated sulphuric acid (50 ml. to 1 l. of carbon tetrachloride) until no colour develops in the acid layer on standing for 2 hr. The carbon tetrachloride is separated, washed with distilled water, potassium hydroxide (two lots of 50 ml. each, 50 %), partially dried by standing overnight in an Erlenmeyer flask with a layer of potassium hydroxide pellets and distilled. The distillate can be used in the determination of iodine numbers, but for the above estimation further drying by constant shaking for several hours with phosphorus pentoxide (50 g. to 1 l. of carbon tetrachloride is necessary). The carbon tetrachloride is filtered into a dry distillation flask protected from atmospheric moisture by a calcium chloride guard tube and distilled from phosphorus pentoxide (10 g. to 1 l. of carbon tetrachloride).

Acetic acid. Reducing substances are removed from the acetic acid by boiling with chromic acid (2 % by weight) under a reflux condenser for 2 hr. The acid is distilled using an efficient

[1] M. G. Lambou and F. G. Dollear, *loc. cit.* (p. 30, n. 3).

splash head to prevent any chromic oxide being carried over.
The first fraction is discarded and the middle fraction is tested
for reducing substances by diluting (2 ml.) with water (10 ml.)
and adding potassium permanganate (0·1 ml., 0·1N). The
pink colour should not be discharged entirely at the end of
2 hr.[1]

Water is removed by heating the acetic acid with acetic
anhydride (5 % by volume) under a reflux condenser for 4 hr.
The acetic anhydride must be free from reducing substances.
It can be purified usually by distillation through a Widmer
column. The first fraction is discarded and the middle fraction is
tested as above after hydrolysing (2 ml.) with water (10 ml.).

Potassium iodide. Potassium iodide is dried at 60° C. and
powdered to pass a 60-mesh sieve.

Thiocyanogen reagent. A solution of bromine is prepared by
measuring the reagent (5·15 ml., A.R.) from a microburette
(10 ml.) into a mixture of acetic acid (250 ml., see above) and
carbon tetrachloride (250 ml., see above). Lead thiocyanate
(50 g., see above) is weighed into a bottle (2 l.) containing acetic
acid (500 ml., see above). The bottle is fastened securely in
a mechanical shaker and the bromine solution is added slowly
(in approximately 10 ml. lots at first and then in larger amounts)
with shaking after each addition until the colour of bromine
disappears. The mixture is shaken during 5 min. after the final
addition of the bromine solution. The precipitate of lead
bromide is allowed to settle, and the liquid is decanted and
filtered through two filter papers into a dry Büchner flask. The
filtrate is filtered into a second dry Büchner flask but still using
the same funnel and papers. The reagent now should be a clear
straw-coloured liquid. The reagent is poured into a bottle (using
a wide funnel) which is sealed against moisture and stored at
5° C.

Throughout the preparation and during storage the reagent
must not be exposed to direct sunlight and moisture must
always be excluded. Immediately before an estimation the

[1] 'Committee on Guaranteed Reagents, American Chemical Society';
W. D. Collins, H. V. Farr, J. Rosin, G. C. Spencer and E. Wichers, *J. Ind. Eng.
Chem.* 1926, **18**, 636–9.

solution is left to stand at room temperature long enough to reach equilibrium.

Glassware. All glassware must be dried at 110° C. during 2 hr. and left to cool in a desiccator.

ESTIMATION BY PERACIDS

$$\text{>C=C<} + R.CO.O.OH \longrightarrow \text{>C——C<}_{O}$$

Perbenzoic acid was used originally for the preparation of hydroxybenzoates and epoxides,[1] but as the reaction was shown to be rapid and quantitative[2] it was adapted to the estimation of unsaturation. Other peracids have been examined including peracetic acid,[3] percamphoric acid,[4] perphthalic acid,[5] and perfuroic acid,[6] but generally perbenzoic acid, and to a lesser degree percamphoric acid, are used as they are readily available. An excess of the reagent is added to a chloroform solution of the olefine, and after addition is completed the excess is found iodometrically. The time of reaction depends on the reagent and also on the reactivity of the unsaturated group. Perfuroic acid is far more reactive than either perbenzoic or percamphoric acids, while perphthalic acid fails completely to add to many unsaturated compounds. The reactivity of an ethylenic bond is influenced usually by the adjacent groupings.[7] Numerous anomalies have been recorded with peracetic and perbenzoic acids; low values are found with *l*-limonene, *cyclo*hexene, anethole and eugenol using peracetic acid,[8] and with *d*-Δ³-

---

[1] N. Prileschajew, *Ber.* 1909, **42**, 4811–15.

[2] R. Pummerer and P. A. Burkard, *ibid.* 1922, **55**, 3458–72; R. Pummerer and F. J. Mann, *ibid.* 1929, **62**, 2636–47.

[3] W. C. Smit, *Rec. trav. chim.* 1930, **49**, 691–6; B. A. Arbuzov and B. M. Mikhailov, *J. prakt. Chem.* 1930, **127**, 92–102.

[4] N. A. Milas and I. C. Cliff, *J. Amer. Chem. Soc.* 1933, **55**, 352–5.

[5] H. Böhme and G. Steinke, *Ber.* 1937, **70**, 1709–13.

[6] Cf. N. A. Milas and A. McAlevy, *J. Amer. Chem. Soc.* 1933, **55**, footnote on p. 352.

[7] H. Meerwein, *J. prakt. Chem.* 1926, **113**, 9–29; J. Böeseken, *Rec. trav. chim.* 1926, **45**, 838–44; J. Böeseken and J. S. P. Blumberger, *ibid.* 1925, **44**, 90–5; W. C. Smit, *loc. cit.* (p. 33, n. 3); G. Charrier and A. Moggi, *Gazz. chim. ital.* 1928, **57**, 736–41).

[8] B. A. Arbuzov and B. M. Mikhailov, *loc. cit.* (p. 33, n. 3).

carene,[1] carotenoids,[2] and $\alpha\beta$-unsaturated aldehydes[3] using perbenzoic acid. It has been shown that the reactivity of an ethylenic bond towards peracids either is reduced or prevented by a carbonyl, a carboxyl or an ester grouping in the $\alpha\beta$ position.[4]

The method has been modified and used to distinguish between the terminal and internal double bonds of polymers.[5] This is possible owing to the greater reactivity of the internal double bonds, although there is difficulty in defining the experimental conditions for eliminating side-chain substitution reactions and for separating the contributions of the vinyl and internal double bonds. An empirical method was used originally which assumed equivalent addition rates for compounds of low molecular weight and for large polymers.[6] More recently an extrapolation method based on the difference in addition rates of the two types of linkage has been described and gives satisfactory results for polymers in which at least 70 % of the double bonds are in the chain.[7]

The olefine (about 10 m.equiv. of double bond) is weighed into a volumetric flask (100 ml.) and dissolved in chloroform (50 ml.). Perbenzoic acid (25 ml., 0·5 N) in chloroform is added, the mixture is diluted (to 100 ml.) with chloroform and then left to stand for half an hour. The excess is determined by taking a measured volume (10 ml.) and adding it to a mixture of potassium iodide (25 ml., 10 %) and acetic acid (25 ml., 0·4 N) in a glass-stoppered bottle. The liberated iodine is titrated with a standard solution of sodium thiosulphate (0·05 N), adding starch (1 ml., 1 %) near the end-point.

A blank determination should be done in parallel.

The reaction can be carried out conveniently at room temperature unless it is intended to distinguish between internal and terminal double bonds when the estimation is done at $6 \pm 1°$ C.

[1] B. A. Arbuzov and B. M. Mikhailov, *J. prakt. Chem.* 1930, **127**, 92–102.
[2] R. Pummerer, L. Rebmann and W. Reindel, *Ber.* 1929, **62**, 1411–18.
[3] K. Bodendorf, *Arch. Pharm.* 1930, **268**, 491–9.
[4] K. Bodendorf, *ibid.*
[5] A. Saffer and B. L. Johnson, *Ind. Eng. Chem.* 1948, **40**, 538–41; E. R. Weidlin, Jr., *Chem. Eng. News*, 1946, **24**, 771–4.
[6] *Ibid.*
[7] A. Saffer and B. L. Johnson, *loc. cit.* (p. 34, n. 5).

PREPARATIONS: *Perbenzoic acid in chloroform.* See G. Braun, *Org. Synth.* col. vol. 1, 1st ed. pp. 422–5, 2nd ed. pp. 431–4. *Percamphoric acid in chloroform.* See N. A. Milas and A. McAlevy, *J. Amer. Chem. Soc.* 1933, **55**, 349–52.

MISCELLANEOUS

(1) The iodine number of fats and oils can be found by Margosches' method.[1] An ethereal, or alcoholic solution of the fat or oil is shaken with an excess of iodine (0·2N, at least 65 % excess) and water (200 ml.) for 1 min. and then left to stand for a further 5–8 min. The excess of iodine is titrated with a standard solution of sodium thiosulphate.

The values generally agree with those found by the methods of Hanus and Hübl.

(2) Kaufmann's method[2] for determining unsaturation is based on the reaction of the olefine with excess of a solution of bromine in absolute methyl alcohol saturated with sodium bromide and titrating the excess with a standard arsenite solution. The results generally agree with those found by the methods of Hanus and Hübl and they are more reliable for several of the $\alpha\beta$-unsaturated acids.[3]

A solution of bromine in methyl alcohol (25 ml., 5·2 ml. of bromine made up to 1 l. with absolute methyl alcohol containing 120–150 parts of sodium bromide in 1000 parts of alcohol) is added to the fat, or oil (0·1–1·0 g. depending on the bromine number) dissolved in chloroform and the mixture is left to stand for $\frac{1}{2}$–2 hr. Concentrated hydrochloric acid (3 ml.) and water (10 ml.) are added and the mixture is titrated with a standard solution of arsenite (4·9 g. of arsenious oxide is dissolved in the minimum amount of potassium hydroxide, about 2 g., and made up to 1 l. with water). Methyl red (0·5 g.) is

[1] B. M. Margosches, W. Hinner and L. Friedmann, *Z. angew. Chem.* 1924, **37**, 334–7; W. Czerny, *Z. deut. Oel-Fett.-Ind.* 1924, **44**, 605; M. Naphtali, *ibid.* 1925, **45**, 77–9; E. Stock, *Farben-Ztg.* 1925, **31**, 403–4; S. Yushkevich, *Masloboino-Khirovoc Delo*, 1928, no. 9, 22–6.
[2] H. P. Kaufmann, *Fette u. Seifen*, 1940, **47**, 4–5; *Untersuch. Lebensm.* 1926, **51**, 3–14.
[3] P. Savary, *Bull. soc. chim. France*, 1950, 624–7.

added to the solution. The arsenite solution is standardized against a potassium bromide-potassium bromate solution $(0 \cdot 1 \text{N})$.

(3) Hübl's method has been used to find the iodine number of a very large number of oils, but now it has been replaced almost entirely by quicker methods (see pp. 20–7).[1] A solution of the fat or oil in chloroform is left to stand with a large excess of an alcoholic solution of iodine containing mercuric chloride. The reagent is probably iodine monochloride formed by the reaction

$$HgCl_2 + 2I_2 = HgI_2 + 2ICl$$

The reaction mixture is left to stand at room temperature in the dark for 8–15 hr., potassium iodide and water are added and the liberated iodine is titrated with a standard solution of sodium thiosulphate. If red mercuric iodide is precipitated on dilution more potassium iodide is added.

(4) A micro-determination of the iodine number has been described.[2] Iodine monobromide dissolved in carbon tetrachloride is added to the unsaturated compound and the reaction is carried out at $0°$ C. Samples of the reaction mixture are titrated at definite intervals to determine the excess of the reagent and the amount of halogen acid formed.

(5) Dienes (3 g.) can be estimated by heating with a freshly prepared toluene solution of maleic anhydride (25 ml., 60 %) in an all-glass apparatus. The mixture is cooled, ether (5 ml.) and water (20 ml.) are added down the condenser and the liquid is then transferred to a separating funnel. The flask is rinsed with ether (20 ml.) and the washings, together with water (25 ml.), added to the funnel. The mixture is shaken, and the ether layer is extracted twice with water (25 and 10 ml. lots). The aqueous extracts are combined and titrated with a standard solution of sodium hydroxide (N) using phenolphthalein as indicator. A blank determination must be done in parallel.[3]

[1] *Dingl. Polyt. Journ.* 1884, **253**, 281; *J. Soc. Chem. Ind.* 1884, **3**, 641–3.

[2] J. O. Ralls, *J. Amer. Chem. Soc.* 1934, **56**, 121–3; cf. *ibid.* 1933, **55**, 2083–94.

[3] B. A. Ellis and R. A. Jones, *Analyst*, 1936, **61**, 812–16; H. P. Kaufmann and J. Baltes, *Fette u. Seifen*, 1936, **43**, 93–7; G. D'yachkov and M. Ermolova, *Caoutchouc and Rubber* (U.S.S.R.), 1937, no. 3, 24.

(6) Some fats and oils, including some of the conjugated products of linoleic acid, do not appear to react with maleic anhydride under the conditions described in para. (5) above. If, however, a small trace of iodine (1 ml., 0·1 % solution in acetone) is added to the reaction mixture an addition product is formed and the excess of maleic anhydride is estimated as above.[1] It has been proposed that the value obtained shall be known as the pandiene number as usually it differs from the value with maleic anhydride alone when a comparable reaction takes place.

(7) Olefines react with nitrogen tetroxide forming oily addition products called nitrosates. The change in volume which takes place on shaking the reaction mixture is a measure of the olefinic unsaturation.[2]

(8) Hydrocarbon mixtures can be analysed quantitatively by fractionating into small samples and determining the density of these fractions before and after removing the olefines. The accuracy of the method has been checked using carefully prepared synthetic mixtures.[3]

GASEOUS OLEFINES

HYDROGENATION

$$\text{>C=C<} + H_2 = \text{>CH—CH<}$$

Gaseous olefines have been estimated by absorption in a suitable liquid. Many reagents have been used including sulphuric acid of varying strengths with or without catalysts, bromine and bromine water, and aqueous solutions of dichromates, potassium permanganate and β-naphthol. These methods give variable results, and it has been found that catalytic hydrogenation is the only rapid method giving a reliable indication of the degree of unsaturation.[4]

[1] J. D. von Mikusch, Z. anal. Chem. 1950, 130, 412–14.
[2] G. R. Bond, Jr., Ind. Eng. Chem. (Anal. ed.), 1946, 18, 692–6; E. T. Scafe, J. Herman, G. R. Bond, Jr., et al. Anal. Chem. 1947, 19, 71–5.
[3] H. Tropsch and W. J. Mattox, Ind. Eng. Chem. (Anal. ed.), 1934, 6, 235–41.
[4] W. A. McMillan, H. A. Cole and A. V. Ritchie, Ind. Eng. Chem. (Anal. ed.), 1936, 8, 105–7.

A suitable apparatus is made from two complete Bureau of Standards water-jacketed, burette-and-compensator assemblies joined together with a catalyst tube.[1] A modification has been suggested recently in which the catalyst tube is surrounded by a boiling water jacket.[2] This makes it possible to measure the unsaturation of certain complex C_5 cuts from low-temperature fractional distillations.

A preliminary hydrogenation is done to establish equilibrium between the catalyst and the residual gases, and occasionally it is necessary to repeat the experiment to saturate the catalyst fully. The procedure as described originally passed hydrogen over the catalyst before each determination, but this is unnecessary between estimations on the same sample, and it may be advantageous to omit this step as equilibrium may not be established fully before the first analysis.

The method assumes that the lower hydrocarbons are ideal gases, but these compounds are known to show considerable deviations from the ideal state. Hydrogen can be assumed to obey the gas laws within the limits of analytical measurements, but if a high degree of accuracy is required corrections should be applied for deviations of the olefinic sample and the saturated hydrocarbon mixed with hydrogen.[3]

An active catalyst is prepared (see below) and protected from poisoning by filling the apparatus with hydrogen from T_3 to T_4 (fig. 7). The mercury in the reservoirs, R_1 and R_2, and in the burettes, B_1 and B_2, is adjusted to the same level, and the gases from the manometers, M_1 and M_2, are drawn into B_1 and B_2 respectively and then discarded to the air. A cylinder of pure hydrogen is connected to T_5 through a reducing valve (see p. 10) and, after clearing out the gases, T_5 is opened to the burette, and hydrogen (70–80 ml.) is drawn into B_1. T_5 is closed and T_3 is opened to C, M_2 and T_4, then T_3 is opened carefully to level M_2. The rate of flow of gas can be controlled by regulating T_1, and when the mercury almost touches the electrical contact, T_1 is closed partly,

[1] M. Shepherd, *Bur. Standards J. Research*, 1931, **6**, 121–67.
[2] J. H. Shively, F. Philgreen and H. Levin, *Anal. Chem.* 1949, **21**, 1566–7.
[3] R. F. Robey and C. E. Morrell, *Ind. Eng. Chem.* (Anal. ed.), 1942, **14**, 880–3.

to give a very slow rate of flow. Contact is made and shown by the right lamp flashing, T_1 is closed quickly and then T_8 is closed to M_2 but left open to C and T_4. M_2 usually is left filled with mercury, and after T_8 is closed to M_2 the hydrogen remaining can be flushed out of T_4. T_3 is then reversed and the mercury brought to T_5' as before.

Hydrogen is again passed into the apparatus through T_5 (about 50 ml., this should be approximately twice the volume of hydrogen needed for the reduction). T_5 is closed, T_3 is opened to the manifold and, with the levels in R_1 and B_1 approximately the same, T_8 is opened to T_3, M_2 and C, and the mercury in M_2 is brought to the level for contact by adjusting R_1 (the balance must always be obtained by letting the mercury flow upwards towards the platinum point and never downwards). T_1' is closed quickly and the reading taken for the hydrogen added.

The olefine (about 50 ml.) is taken into B_2 in a similar way except that T_8 is opened to T_4 before the latter is reversed. M_2 is set, T_2 is closed quickly and the volume of the sample is noted.

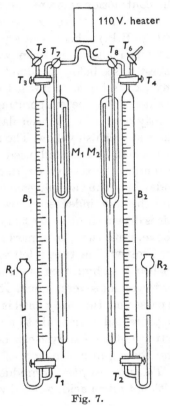

Fig. 7.

T_8 is opened and the mercury in M_2 is pulled up until it just touches the tap which is then closed to the manometer. This is an important step and should not be neglected. T_1 and T_2 are opened fully, the hydrogen is passed from B_1 to B_2 and the gases are mixed thoroughly by alternately raising and lowering R_1 and R_2 four or five times. The taps are then closed slightly so that it takes about 2 min. to pass the gas from one burette to the other. After five complete cycles the mercury in B_2 is raised to T_4

which is closed, R_1 and B_1 are balanced approximately, T_8 is opened carefully under a little pressure to the manifold and T_4, the mercury level, is set to the point in M_2 and the contraction is found from the reading in B_1.

Two complete cycles are made with the mixture of gases until the contraction remains constant. The reading is then taken.

PREPARATIONS: *The catalyst.* Nickel nitrate hexahydrate (4·0 g., B.P.) is heated in a porcelain evaporating basin (50 ml.) until it melts in its own water of crystallization. Shredded asbestos (see below) is added until all the solution is absorbed; any excess liquid is pressed out with a porcelain spatula and drained from the evaporating basin. The asbestos is calcined lightly over a low bunsen flame until some appears black and the rest greenish yellow. The mass is left to cool, then broken up into small pieces and used to fill a U-shaped catalyst tube (8·0 mm., Pyrex) to a depth of about 6·5 cm. on each side. The ends of the tube are plugged with ignited asbestos, or glass wool, and the tube indented below the plugs to keep them in place. Air is drawn through the tube which is heated to 310° C., but not above, until the whole mass is black and no more brown fumes are evolved. The tube is left to cool, connected to the apparatus, flushed with hydrogen and then heated at 325° C. in a stream of hydrogen for several hours. It is left to cool finally and hydrogen is passed continuously until it is cold. The catalyst should always be protected against air to avoid poisoning, but if necessary the activity of the catalyst can be restored by reheating at 325° C. and cooling in hydrogen.

The asbestos fibre. Shredded, long-fibre asbestos is digested with hot nitric acid, washed well with water, dried and ignited.

CHAPTER II

COMPOUNDS CONTAINING THE HYDROXYL GROUP

ALCOHOLS

Reaction with acetic anhydride, acetyl chloride or acetic acid to form an ester is the simplest volumetric method for the estimation of alcohols. The method often fails with tertiary alcohols, as olefines are formed under the experimental conditions necessary for quantitative esterification, and even with primary and secondary alcohols there are great differences in reactivity of the hydroxyl group. An alternative method which is used frequently measures the volume of methane formed in the reaction between the reactive hydrogen of the alcohol and methyl magnesium iodide. The determination has been modified recently to use lithium aluminium hydride as the reagent. Alcohols can be estimated by the following general methods:

Esterification:

A. Acetylation with acetic anhydride:
$$R.OH + (CH_3.CO)_2O = CH_3.COOR + CH_3.COOH.$$

B. Acetylation with acetyl chloride:
$$R.OH + CH_3.COCl = CH_3.COOR + HCl.$$

C. Acetylation with acetic acid:
$$R.OH + CH_3.COOH = CH_3.COOR + H_2O.$$

Measurement of gaseous reaction products:

A. Methane from methyl magnesium iodide—Zerewitinoff's method:
$$R.OH + CH_3.Mg.I = CH_4 + Mg{\overset{\displaystyle I}{\underset{\displaystyle OR}{\diagdown\!\!\!\diagup}}}.$$

B. Hydrogen from lithium aluminium hydride:
$$4R.OH + Li.Al.H_4 = 4H_2 + LiOR + Al(OR)_3.$$

Colorimetric.
Miscellaneous.
Specific for methyl alcohol.

ESTERIFICATION

A. Acetylation with Acetic Anhydride

$$R.OH + (CH_3.CO)_2O = CH_3.COOR + CH_3.COOH.$$

Alcohols can be estimated simply by determining the amount of acetic anhydride used in the formation of the acetyl derivative.[1] The estimation was carried out originally by heating a mixture of the alcohol with an excess of acetic anhydride and determining the excess after the completion of the reaction. The acetic acid formed in the esterification was neutralized, the unchanged acetic anhydride and the ester were hydrolysed with excess standard alkali and the excess was back-titrated with acid. In this procedure the reagent must be measured with great accuracy, as 0·1 ml. of acetic anhydride is equivalent to 18·1 ml. of a decinormal solution of sodium hydroxide. The reaction is tedious frequently, as elaborate precautions are needed for the analysis of a volatile alcohol while the rate of reaction is very slow with the higher alcohols. Many modifications have been described and the methods now used generally overcome these difficulties by using a mixture of acetic anhydride and pyridine as the acetylating reagent. Some pyridine acetate is formed, but neither the acetate nor the excess pyridine interfere in the final titration if phenolphthalein or a mixture of cresol red and thymol blue is used as the indicator. The end-point may be difficult to determine at first with phenolphthalein as fading occurs frequently, but one or two titrations usually give the experience needed for accurate results. Indicators are not satisfactory if the reaction mixture is coloured; the end-point should then be found potentiometrically.[2]

[1] A. Verley and Fr. Bolsing, *Ber.* 1901, **34**, 3354–8; E. André, *Fettchem. Umschau*, 1933, **40**, 194–7; V. L. Peterson and E. S. West, *J. Biol. Chem.* 1927, **74**, 379–83; E. S. West, C. L. Hoagland and G. H. Curtis, *ibid.* 1934, **104**, 627–34; S. Marks and R. S. Morrell, *Analyst*, 1931, **56**, 428–9.
[2] C. L. Ogg, W. L. Porter and C. O. Willits, *Ind. Eng. Chem.* (Anal. ed.), 1945, **17**, 394–7; A. Robertson and W. A. Waters, *J. Chem. Soc.* 1948, pp. 1585–90.

The method has been adapted to the micro-estimation of alcohols and phenols.[1] The procedure may be similar to the one used on the macro-scale (see below), or the reaction mixture can be sealed in a tube and left at room temperature overnight or as long as is necessary for complete esterification.[2]

Primary and secondary alcohols can be estimated by this method, but tertiary alcohols undergo partial or complete dehydration and the water so formed reacts with the reagent to give two equivalents of acetic acid. Phenols, primary and secondary amines and amino compounds, mercaptans, aldehydes and other compounds which react with acetic anhydride must be absent. Low values may result either from the loss of a volatile alcohol during the esterification (e.g. the lower aliphatic alcohols) or from incomplete acetylation (e.g. geraniol 90 %, terpineol 3 %, benzyl alcohol 92 %).

A mixture of the alcohol (0·1–0·3 g.) and the reagent (about 2 g., see below) is heated in a boiling water-bath under an efficient reflux condenser for ¼–2 hr. The solution is left to cool, washed into a beaker with ice-cold water and titrated rapidly with a standard solution of sodium hydroxide (0·5N), using phenolphthalein as the indicator. A blank determination is carried out in parallel.

When the alcohol decomposes below 100° C. the reaction can be carried out, without loss of accuracy, by heating the mixture at 60–80° C. for 24–48 hr.

1. *Potentiometric titration*

A glass-stoppered iodine flask is modified by sealing into it two side-arms which can be closed with glass stoppers or be used for the two electrodes.

The acetylating reagent of acetic acid and pyridine (3 ml., see below) is added to the alcohol (1–2·5 m.equiv.) from a micro-burette (about 5 ml.) with the reservoir protected by a drying tube. The centre stopper of the reaction vessel is moistened with pyridine and placed lightly in the flask to allow for expansion on

[1] F. H. Stodola, *Mikrochemie*, 1937, 21, 180–3.
[2] J. W. Petersen, K. W. Hedberg and B. E. Christensen, *Ind. Eng. Chem.* (Anal. ed.), 1943, 15, 225–6.

heating. The mixture is heated on a boiling water-bath during 45 min., water (5–6 ml.) is added to the well at the top of the flask and the stopper is removed slowly so that the walls of the flask are washed. The mixture is heated again during 2 min. and then cooled with the centre stopper partly open. The three stoppers and the sides of the flask are washed with n-butyl alcohol (10 ml., technical may be used), glass and calomel electrodes are inserted through the side-arms and the acid is titrated with an alcoholic solution of sodium hydroxide (0·5 or 0·1 N, see below) to pH 9·8. The free acid in the sample is found by repeating the titration using pyridine only instead of the acetylating mixture and adding neutral ethyl alcohol (5 ml.) just before the titration to make the solution homogeneous.

If the reaction mixture is colourless a mixed indicator of cresol red and thymol blue can be used (see below).

Percentage of hydroxyl

$$= \frac{\text{ml. of reagent (sample} - \text{blank)} \times N \times 100 \times 17}{1000 \times \text{wt. of alcohol in g.}}.$$

Accuracy $\pm 0\cdot2$ $(1\cdot0)\%$.

PREPARATIONS: *Acetylating reagent.* Glacial acetic acid (1 part, B.P.) is added to pure, dry pyridine (3 parts, see p. 62). Whenever possible the reagent is made up immediately before it is used.

Mixed indicator. An aqueous solution of cresol red (1 part, 0·1 %, neutralized with sodium hydroxide) is added to an aqueous solution of thymol blue (3 parts, 0·1 %, neutralized with sodium hydroxide).

Sodium hydroxide. The alcoholic solution of sodium hydroxide is prepared by adding aldehyde-free alcohol (cf. p. 188) to a saturated solution of sodium hydroxide (about 18 N).

2. *Micro-estimation*

The alcohol (5–20 mg.) and the acetylating mixture (at least twice the theoretical amount) are heated at 95–100° C. under a reflux condenser for 1 hr. The estimation is then carried out as above, titrating the acid with a standard solution of sodium hydroxide in alcohol (0·02 N, free from carbon dioxide). A blank determination is carried out in parallel.

PREPARATION: *The acetylating mixture.* Glacial acetic acid (1 part, B.P.) is added to pure, dry pyridine (4 parts, see p. 62).

B. Acetylation with Acetyl Chloride

$$R.OH + CH_3.COCl = CH_3.COOR + HCl.$$

Acetyl chloride, under the same conditions of temperature and concentration, is considerably more reactive than either acetic anhydride or a mixture of acetic anhydride and acetic acid. Its use has been restricted until recently to qualitative reactions owing to the difficulty of measuring and using small quantities of this easily hydrolysable substance without loss. It can now be used successfully either as a paste of acetyl pyridinium chloride in dry toluene[1] or by measuring out the acetyl chloride in a special pipette.[2] The values so determined compare very favourably with those found using an acetylation mixture of acetic anhydride and pyridine (see pp. 42–5). The alcohol and an excess of the reagent are mixed and left to stand until the reaction is finished; the excess of acetyl chloride is hydrolysed and estimated by titration with a standard solution of alkali. Fading of the indicator takes place occasionally in this titration, but when it is not too rapid accurate results are possible by adding small, constant volumes of the alkali near the end-point until the colour persists for half a minute. A blank determination should be done at the same time.

The method gives accurate values for a wide range of alcohols, including primary and secondary aliphatic and aromatic alcohols and polyhydroxy compounds which dissolve readily in the solvent. Tertiary alcohols give very low results and cannot be estimated in this way. Mono-substituted glycol ethers give good results when the reaction is done at room temperature, but diethylene glycol gives high values even under these conditions.[3] The hydroxyl content of cinnamyl alcohol

---

[1] D. M. Smith and W. M. D. Bryant, *J. Amer. Chem. Soc.* 1935, **57**, 61–5.

[2] B. E. Christensen, L. P. Pennington and P. K. Dimick, *Ind. Eng. Chem.* (Anal. ed.), 1941, **13**, 821–3; see p. 46.

[3] A. L. Montes, *Anales assoc. quim. Argentina*, 1943, **31**, 109–13.

cannot be determined with acetyl chloride alone, and it gives a low result (97·9 %) with acetyl pyridinium chloride. α- and β-naphthol also give erratic values. It is probable that side reactions take place through the hydrogen chloride formed. The method cannot be used when phenols, primary and secondary amines and amino compounds, mercaptans, the higher fatty acids or any substances which form stable compounds with acetyl chloride are present. Water when present in a large amount competes with the alcohol for the reagent, so that less of the substance to be analysed is taken to keep a suitable excess of the acetylating agent. The sensitivity of the method thereby is reduced and the estimation usually is unsatisfactory. Formides, oxalides, many α-hydroxy esters of low molecular weight and other substances which hydrolyse easily also interfere. Aldehydes and nitroso compounds interfere with the end-point, and aldehydes also react slowly with acetyl chloride, but the effect is less marked than with acetic anhydride and pyridine and is not serious unless more than 10 % is present. Ketones have little effect, and any possible error is cancelled in a blank estimation.

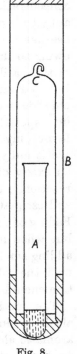

Fig. 8.

1. *Using a special pipette*

The alcohol (0·1–0·7 g., depending on the percentage of hydroxyl) is weighed into the tube A (fig. 8) (about 10 cm. in length; the tube is closed with a cork during weighing if the alcohol is volatile) and cooled in carbon dioxide. Acetyl chloride (0·5 ml., an excess of about 50 %) is added from the pipette (see below) and the tube is placed in a larger tube B (about 20 cm. in length) containing water (5–10 ml.). The apparatus is completed by covering A with an inverted tube C (about 15 cm. in length) and closing B with a stopper. The mixture is heated in a water-bath kept at $40 \pm 1°$ C. during 20 min. The excess acetyl chloride is hydrolysed by inverting the vessel and mixing the liquids. The mixture is cooled and titrated rapidly with a standard solution of sodium hydroxide ($0\cdot1$ N) using phenolphthalein as the indicator. A blank determination should be carried out in parallel.

Percentage of hydroxyl

$$= \frac{(\text{ml. of NaOH in blank} - \text{ml. of NaOH in estimation}) \times \text{N} \times 17}{10 \times \text{wt. of alcohol (in g.)}}.$$

Accuracy $\pm\, 0\cdot5$–$1\cdot0\,\%$.

Construction and use of the pipette. The pipette (fig. 9) is made from 10 mm. Pyrex tube, is about 21 cm. long, and delivers about $0\cdot5$ ml. of liquid. The constriction is $0\cdot1$ mm. The rubber cap is split longitudinally at the end so that it can be filled and emptied by pressing parallel, or at right angles to the slit.[1]

Fig. 9.

An accurate volume is measured by filling the pipette and draining the tube to the constriction. The liquid is transferred to the reaction vessel by opening the slit, allowing the liquid to drain, and then pressing the closed cap to force all the liquid from the lower capillary.

2. *Using acetyl pyridinium chloride in dry toluene*

Acetyl chloride (10 ml., see below) in dry toluene is measured into a long-necked flask from a Lowry automatic pipette. The solution is cooled in ice for 1 min., pure, dry pyridine (2 ml., see p. 62) is added, a stopper is placed in the flask and the mixture is shaken vigorously until a thin paste of acetyl pyridinium chloride is formed. A known weight of the dry alcohol is added, so that for each hydroxyl group reacting an excess (50 %) of the reagent remains. The flask is placed in a water-bath kept at $60 \pm 1^\circ$ C., the stopper is loosened momentarily to release the pressure and the mixture is heated with occasional shaking during 20 min. The mixture is cooled in ice, water (25 ml., ice-cold) is added and the excess acetyl pyridinium

[1] Cf. K. Linderström-Lang and H. Holter, *Compt. rend. trav. lab. Carlsberg,* 1931, **19**, no. 4; *Z. physiol. Chem.* 1931, **201**, 9–30.

chloride is hydrolysed. The reagent remains dissolved in the toluene layer, and vigorous shaking is needed to complete the hydrolysis. The mixture is cooled and titrated rapidly with a standard solution of sodium hydroxide using phenolphthalein as the indicator. A blank determination should be carried out in parallel.

Accuracy ± 0.5 (2·0) %.

PREPARATION: *Acetyl chloride in dry toluene.* Acetyl chloride (117·75 g., 1·5 mol.) is dissolved in dry toluene and the solution is made up (to 1 l.) with the same solvent.

C. Acetylation with Acetic Acid

$$R.OH + CH_3.COOH = CH_3.COOR + H_2O.$$

The determination of the water formed in the complete esterification of a hydroxyl group has been used as a method of estimation for alcohols.[1] The acetate is prepared and the equilibrium is adjusted by using a large excess of acetic acid to favour the ester and water formation almost entirely. Equilibrium is reached quickly using boron trifluoride as a catalyst, and the water is estimated rapidly and accurately by titration with the Karl Fischer reagent (see below; cf. pp. 150, 173, 175, 197). Blank titrations must be carried out to determine the amount of water in the solvent, in the catalyst, in the reagent and in the alcohol.

The method can be applied to aliphatic, aromatic and alicyclic alcohols. Aliphatic tertiary alcohols can be estimated accurately as an equivalent amount of water is formed in the reaction with acetic acid either by esterification or by dehydration (cf. pp. 43, 45). It is unnecessary that the alcohol should be dry as the error can be eliminated by a blank determination; an aqueous solution containing as little as 0·8 % of ethyl alcohol has been analysed successfully. It is claimed that aliphatic alcohols can be estimated approximately in the presence of aromatic alcohols and phenols by using different concentrations of the catalyst.[2]

[1] W. M. D. Bryant, J. Mitchell, Jr., and D. M. Smith, *J. Amer. Chem. Soc.* 1940, **62**, 1–3.
[2] *Ibid.*

A number of substances interfere as they react with acetic acid to form water under the conditions used for esterification. Aldehydes and ketones react to a marked degree (e.g. 37 % aqueous formaldehyde, 70 %; $cyclo$hexanone, 50 %) and acetals and ketals form almost the theoretical amount of water (2 mols; methylal, 92 %; acetal, 91 %; di-n-butyl formal, 88 %):

$$\begin{array}{c} R \\ \diagdown \\ C{=}O + 2CH_3.COOH = \\ R' \diagup \end{array} \begin{array}{c} R \diagdown \quad \diagup O.COCH_3 \\ C \\ R' \diagup \quad \diagdown O.COCH_3 \end{array} + H_2O$$

$$CH_2(OCH_3)_2 + 4CH_3.COOH = 2CH_3.COOCH_3 + (CH_3.COO)_2CH_2 + 2H_2O.$$

High values are obtained with a number of unsaturated terpene alcohols (e.g. geraniol, 109 %; terpineol, 114 %) and low values generally result from a reduction in the activity of the catalyst when amines are present.

The alcohol (5–10 g.) is weighed into a graduated flask about one-third filled with dioxane, and a standard solution is made up by adding more of the solvent. The solution is shaken until it is homogeneous and the flask is placed in a thermostatically regulated water-bath ($25 \pm 1°$) for half an hour. A sample is withdrawn (5 ml.) with a pipette, transferred to a flask (250 ml.) and a solution of boron trifluoride in acetic acid (20 ml., see below) is added. The flask is stoppered tightly and heated in a thermostatically controlled oven or water-bath at $67 \pm 2°$ C. for 2 hr. The mixture is allowed to cool, pyridine is added (5 ml.) and the contents of the flask are titrated with the Karl Fischer reagent (see pp. 50–5).

A blank determination must be carried out under identical conditions. A mixture of dioxane (5 ml.) and boron trifluoride in acetic acid (20 ml.) is used. The increase in the amount of water in the sample over the blank is a direct measure of the percentage of alcohol present. If the alcohol is not anhydrous the percentage of water must be found by titrating a weighed amount with the Karl Fischer reagent.

Percentage of hydroxyl

$$= \frac{\text{ml. of reagent (sample} - \text{blanks)} \times \text{normality} \times 17}{100 \times \text{wt. of alcohol}}.$$

Accuracy ± 1 %.

PREPARATIONS: *Boron trifluoride catalyst.*[1] Boron trifluoride
gas (about 100 g.) is dissolved in glacial acetic acid (250 ml.)
and water (1–2 ml.), and the resulting solution is made up to
1 l. by the addition of glacial acetic acid.

The Karl Fischer reagent. The simplest and most widely
applicable method for estimating small amounts of water uses
a specific reagent prepared by the action of sulphur dioxide on
iodine dissolved in methyl alcohol and pyridine.[2] The reaction
in aqueous solution is

$$I_2 + SO_2 + 2H_2O \rightleftharpoons 2HI + H_2SO_4,$$

and originally it was thought that the addition of pyridine only
forced the reaction to the right by removing the acidic products
so that one molecule of iodine should be equivalent to two
molecules of water. This assumption has been disproved, and it
is known that both solvents take part in the reaction and that
one molecule of iodine is equivalent to one molecule of water.[3]
The reaction, in its simplest form, is

$$I_2 + SO_2 + H_2O + 3C_5H_5N \longrightarrow 2C_5H_5N \cdot HI + C_5H_5N\overset{\displaystyle SO_2}{\underset{\displaystyle O}{\big\langle}}|,$$

$$C_5H_5N\overset{\displaystyle SO_2}{\underset{\displaystyle O}{\big\langle}}| + CH_3OH \longrightarrow C_5H_5N\overset{\displaystyle SO_4CH_3}{\underset{\displaystyle H}{\big\langle}}$$

The accuracy of the method depends upon the care with
which the end-point is found. The reagent is made up in dry
solvents which are hygroscopic, and although the original
procedure of direct titration of the substance dissolved in
anhydrous methyl alcohol or dioxane gives fairly concordant
results, the values are high because of absorption of water. This
error is eliminated by using a small enclosed titration vessel
fitted with guard tubes to exclude moisture. The direct titration
with visual determination of the end-point is restricted to
estimations where the reddish brown colour of free iodine can
be seen clearly, but in coloured solutions this difficulty is

[1] Cf. H. Bowlus and J. A. Nieuwlands, *J. Amer. Chem. Soc.* 1931, **53**,
3835–40; H. D. Hinton and J. A. Nieuwlands, *ibid.* 1932, **54**, 2017–18;
G. F. Hennion, H. D. Hinton and J. A. Nieuwlands, *ibid.* 1933, **55**, 2857–60.
[2] K. Fischer, *Angew. Chem.* 1935, **48**, 394–6; cf. R. Bunsen, *Ann.* 1853, **86**,
265–91.
[3] D. M. Smith, W. M. D. Bryant and J. Mitchell, Jr., *J. Amer. Chem. Soc.*
1939, **61**, 2407–12.

overcome by the electrometric detection of the end-point.[1] The addition of an excess of the Karl Fischer reagent followed by a back titration with a standard solution of water has been found to give a sharper and more reproducible end-point than the direct titration to a visual end-point, but the direct and indirect electrometric methods which are described below give precise values, and the same degree of accuracy is possible in the majority of estimations. Both procedures use the principle of the 'dead-stop end-point'.[2] In this method two platinum electrodes are immersed in the reaction mixture and a small e.m.f. is applied so that a current flows, as long as iodine is present, to remove the hydrogen and depolarize the cathode., In the indirect titration therefore the sensitive micro-ammeter in the circuit returns to zero when iodine is eliminated and polarization occurs, whereas in the direct titration the current increases suddenly as soon as there is an excess of the iodine reagent. An alternative way of showing the end-point uses a 'magic eye' electronic indicating tube in a suitable circuit.[3]

(a) Indirect titration. The apparatus is as shown in fig. 10.[4] The Karl Fischer reagent and a standard water solution are stored in the two dark-coloured bottles and are pumped into the respective burettes by hand bellows, the air being dried by passing through a tower packed with calcium chloride. The burettes (10 ml., 0·02 ml. graduations) are closed with tubes filled with calcium chloride and a short length of silica gel at the end; this involves no loss of water from the very dilute standard solution. An accurately ground Bakelite stopper fits into the neck (B 34 ground joint) of the titration vessel (60–70 ml.), and the two electrodes, the jets of the burettes, the stirrer, and an inlet tube for nitrogen pass through the stopper and are sealed in completely with wax. The stopper and the stirrer are coated thickly with soft paraffin, and if desired the latter can be omitted from the apparatus as efficient stirring is possible by bubbling

[1] E. G. Almy, W. E. Griffin and C. S. Wilcox, *Ind. Eng. Chem.* (Anal. ed.), 1940, **12**, 392–6; G. Wernimont and F. J. Hopkinson, *ibid.* 1943, **15**, 272–4.
[2] *Ibid.*; cf. C. W. Foulk and A. T. Bawden, *J. Amer. Chem. Soc.* 1926, **48**, 2045–50.
[3] R. Kieselbach, *Anal. Chem.* 1949, **21**, 1578–9.
[4] T. G. Bonner, *Analyst*, 1946, **71**, 483–9.

dry nitrogen through the liquid in the reaction vessel (cf. the direct method). A tube of silica gel, two drying bottles each containing concentrated sulphuric acid and a tube of anhydrous magnesium perchlorate (anhydrone) is a satisfactory drying train for removing the last traces of moisture from the gas. The

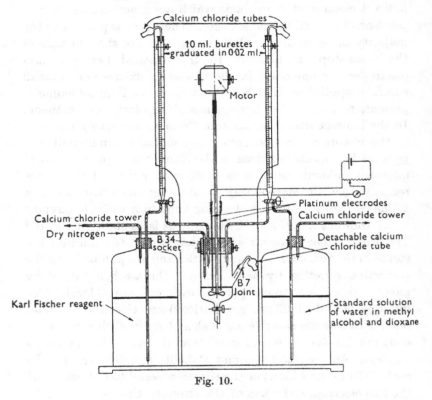

Calcium chloride tubes

10 ml. burettes
graduated in 0·02 ml.

Motor

Platinum electrodes
Calcium chloride tower

Calcium chloride tower
Dry nitrogen

B 34
socket

Detachable calcium
chloride tube

B 7
Joint

Karl Fischer reagent

Standard solution
of water in methyl
alcohol and dioxane

Fig. 10.

solution to be estimated is added to the reaction flask through a side arm closed with a small calcium chloride tube fitted with a ground-glass joint (B 7).

Dry nitrogen should be passed through the apparatus for 15 min. before an estimation. The liquid (a measured volume) and the Karl Fischer reagent (an excess is run in from the burette which is equivalent to at least 1 ml. of the standard water solution) are added to the titration flask, the mixture is stirred and a small e.m.f. is applied so that the galvanometer

ALCOHOLS

needle is off the scale (about 80 mV., although no decrease in
sensitivity is found if 200 mV. or higher is used). The excess
reagent is titrated with the water solution (about 5 mg. of water
per ml.) until the colour of the mixture is yellowish brown, after
which the addition is continued drop by drop until the galvanometer needle 'kicks'. The stirrer is stopped and the end-point
is found by adding the water solution dropwise until the needle
returns to zero. The titration is repeated until three successive
readings agree within 0·01 ml. The liquid should not be stirred
too rapidly, as a violent mechanical action tends to sweep off the
hydrogen which accumulates on the cathode at the end-point.
The water solution must be standardized at least once a week,
and if the titration vessel is detached during a series of estimations it must be restandardized before continuing the analysis.
This is done by taking two glass-stoppered graduated flasks
(50 ml.) adding a weighed amount of water to one (about 0·1 g.),
using a Lunge Rey pipette or by any convenient method, and
then filling both flasks to the mark with dry dioxane. A mixture
of the Karl Fischer reagent (5 ml., from the burette) and dioxane
(2·5 ml.) is titrated with the solution of water to be standardized
(stored in the apparatus). In the same way the dioxane-water
mixture from the other graduated flask is titrated, and the
difference in volume between the two titrations is equal to one-
twentieth of the water added to the dioxane.

(b) Direct titration. The apparatus is as shown in fig. 11.
Moisture is excluded from the apparatus by fitting a guard tube
filled with anhydrous magnesium perchlorate (anhydrone) to the
burette and drying the nitrogen by passing it through a sample
of the Karl Fischer reagent. This dries the gas sufficiently well
to cause little or no decomposition of the reagent when passed
into the reaction vessel at a rate needed for efficient stirring.
At the beginning of the titration the micro-ammeter registers
between 5 and 10 scale divisions, and as soon as an excess of
the iodine reagent is present the deflexion rises to some 80–90
divisions. The general procedure for the titration is similar to
that described below for standardization of the reagent.[1]

[1] *Moisture determination by the Karl Fischer reagent* (British Drug Houses
booklet), p. 9.

Anhydrous methyl alcohol (5 ml.) is placed in the apparatus and titrated with the Karl Fischer reagent until the galvanometer gives an increased deflexion which does not return to its

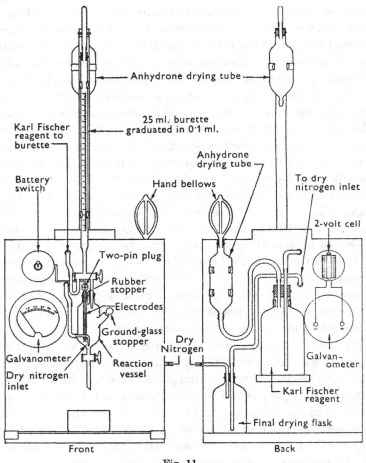

Fig. 11.

original value for at least 30 sec. An accurately prepared solution of water (10 ml., 0·5 %) in anhydrous methyl alcohol is added, the titration is repeated, and the difference between the two values is the volume of the reagent equivalent to the water present (50 mg.).

An alternative method of standardization has been described and uses a hydrated salt of definite composition. Sodium acetate trihydrate[1] is chosen frequently as it is readily soluble in dry methyl alcohol, but the water content of each batch should be found before it is used for standardization in order to prevent any error caused by variation in hydration. This procedure is generally more tedious than using the commercially available standard solution of water in methyl alcohol. Other salts which have been used include ammonium oxalate monohydrate,[2] citric acid monohydrate[3] and recently sodium tartarate dihydrate.[4]

The reagent is prepared[5] by passing sulphur dioxide slowly into a solution of iodine (63 g., A.R.) in dry pyridine (100 ml.) which is cooled in a freezing mixture and stirred continuously until an increase in weight (of 32 g.) is obtained. The mixture is left to stand for 30 min. and then diluted (to 500 ml.) with anhydrous methyl alcohol.

The reagent is not completely stable as the solvents used in its preparation are hygroscopic and, even in the absence of water, reactions take place which at present are not fully understood. The solution should be standardized daily whenever it is in use (see above).

MEASUREMENT OF GASEOUS REACTION PRODUCTS

A. Methane from Methyl Magnesium Iodide:
Zerewitinoff's Method

$$R.OH + CH_3.Mg.I = CH_4 + Mg\begin{matrix} \diagup I \\ \diagdown OR \end{matrix}.$$

Alcohols react quantitatively with methyl magnesium iodide and can be estimated by measuring the volume of methane liberated.[6] The method is used frequently for determining the

---

[1] *Loc. cit.* (p. 53, n. 1); G. G. Warren, *Can. Chem. Process Inds.* 1945, **29**, 370.

[2] R. P. Rennie and J. L. Monkman, *ibid.* 1945, **29**, 370.

[3] G. R. Cornish, *Plastics (London),* 1946, **10**, 99; G. K. Jones, *Paint Mfg,* 1945, **15**, 360.

[4] J. D. Neuss, M. G. O'Brien and H. A. Frediani, *Anal. Chem.* 1951, **23**, 1332–3.

[5] *Moisture determination by the Karl Fischer reagent* (British Drug Houses booklet), p. 4; cf. K. Fischer, *loc. cit.* (p. 50, n. 2); D. M. Smith, W. M. D. Bryant and J. Mitchell, Jr., *loc. cit.* (p. 50, n. 3).

[6] H. Hibbert and J. J. Sudborough, *J. Chem. Soc.* 1904, **85**, 933–8; cf. L. Tschugaeff, *Ber.* 1902, **35**, 3912–14.

percentage of hydroxyl, and accurate results are obtained when the necessary precautions are taken.[1] The method cannot be used when water, phenols, primary and secondary amines and amino compounds, mercaptans, acids, esters, acid halides, cyanides, isocyanides, alkyl halides are present, or any substance which can react with a Grignard reagent.

The solvent generally used in the preparation of the Grignard reagent is di*iso*amyl ether (see p. 61), although anisole has been suggested and is satisfactory. The vapour pressure of these solvents is low (di*iso*amyl ether at 20° C., 2·8 mm.), and they contribute little to the volume of gas which is measured. Diethyl ether should not be used owing to its high vapour pressure (442·2 mm. at 20° C.), although modifications of the usual method are described in which it is the solvent.[2] All unchanged methyl iodide must be removed from the solution of the reagent either by heating under reduced pressure,[3] or when the apparatus shown in fig. 13 is used by heating in an oil-bath at 120° C. and passing pure, dry nitrogen over the reagent. If this is not done high values can result in estimations where a solution of the alcohol in pyridine is used, probably from ethane formed by the following reactions:[4]

$$C_5H_5N + CH_3 . I = C_5H_5N \diagdown \begin{matrix} CH_3 \\ I \end{matrix},$$

$$C_5H_5N \diagup \begin{matrix} CH_3 \\ I \end{matrix} + CH_3 . Mg . I = C_5H_5N + MgI_2 + C_2H_6.$$

The reagent should be standardized and kept dry and only added to the reaction vessel immediately before the estimation.

A solution of the alcohol in di*iso*amyl ether, pyridine, quinoline, anisole, toluene, xylene, petrol ether or dioxane is usually

[1] W. Hollyday and D. L. Cottle, *Ind. Eng. Chem.* (Anal. ed.), 1942, **14**, 774–6; M. Lieff, G. F. Wright and H. Hibbert, *J. Amer. Chem. Soc.* 1939, **61**, 865–7; E. P. Kohler, J. F. Stone and R. C. Fuson, *ibid.* 1927, **49**, 3181–8; A. P. Tanberg, *ibid.* 1914, **36**, 335–7; R. Ciusa, *Gazz. chim. ital.* 1920, [ii], **50**, 53–5; Th. Zerewitinoff, *Ber.* 1907, **40**, 2023–31; *ibid.* 1909, **42**, 4802–8; *ibid.* 1910, **43**, 3590–5; *ibid.* 1912, **45**, 2384–9; *ibid.* 1914, **47**, 1659 & 2417–23; B. Oddo, *ibid.* 1911, **44**, 2048–52.

[2] R. Ciusa, *loc. cit.* (p. 56, n. 1). A. P. Terent'ev and A. I. Kireeva, *Isvest. Akad. Nauk S.S.S.R. Otdel. Khim. Nauk*, 1951, 172–8.

[3] P. M. Marrian and G. F. Marrian, *Biochem. J.* 1930, **24**, 746–52.

[4] Th. Zerewitinoff, *loc. cit.* (p. 56 n. 1); P. M. Marrian and G. F. Marrian, *loc. cit.* (p. 56, n. 3).

used, although estimations have been done without a solvent.[1] The concentration of the alcohol appears to be important, as low results have been recorded with solutions containing more than 0·1 mol. of the alcohol per litre.[2] This value should be the upper limit, and even more dilute solutions should be used whenever possible (about 0·07 mol. per l.). The formation and precipitation of the sparingly soluble alkoxyl magnesium iodide prevents the reaction going to completion and low values result.[3] Very little solid is deposited when the concentration of the alcohol is kept below the value suggested above. Many conflicting statements about solvents appear in the literature, but the lack of agreement may be due to estimations being carried out using highly concentrated solutions of the alcohol. Pyridine, di*iso*amyl ether and anisole are the solvents used most frequently on account of their good solvent power and very low vapour pressure. Solutions of glucose and of benzoic and cinnamic acids in dioxane react incompletely, as do vanillin and *iso*vanillin in both dioxane and di*iso*amyl ether, but in pyridine and xylene the estimations are quantitative.[4]

If air is present in the apparatus oxygen will be absorbed by the reagent and low results will be obtained.[5] This error is eliminated either by replacing the air with an indifferent gas, or by allowing the oxygen in the apparatus to be absorbed completely before proceeding with the estimation. The former is adopted generally and pure, dry nitrogen is usually used, although methane and dimethyl ether have been suggested.[6] When the air is replaced by nitrogen care must be taken to avoid a loss of alcohol which can be very serious in the estimation of volatile alcohols.

The reaction is usually quantitative in pyridine at room temperature but heating is necessary in other solvents, particularly anisole, and when more than one hydroxyl group is

[1] E. P. Kohler, J. F. Stone and R. C. Fuson, *loc. cit.* (p. 56, n. 1).
[2] H. Hibbert and J. J. Sudborough, *loc. cit.* (p. 55, n. 6); B. Oddo, *loc. cit.* (p. 56, n. 1).
[3] M. Lieff, G. F. Wright and H. Hibbert, *loc cit.* (p. 56, n. 1); E. P. Kohler, J. F. Stone and R. C. Fuson, *loc. cit.* (p. 56, n. 1).
[4] M. Lieff, G. F. Wright and H. Hibbert, *loc. cit.* (p. 56, n. 1).
[5] Cf. J. Meisenheimer and W. Schlichenmayer, *Ber.* 1928, **61**, 2029–43.
[6] R. Ciusa, *loc. cit.* (p. 56, n. 2).

present. The reaction vessel should be immersed in a water-bath at 70–100° C. for 10 min. to ensure that the reaction is completed. Longer times of heating and higher temperatures have been recommended but no advantages are gained.

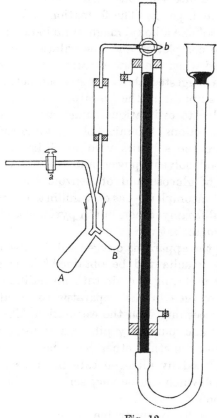

Fig. 12.

The apparatus, alcohol, solvent, and nitrogen must be dried (see pp. 52, 64), and a blank determination should always be carried out before the estimation to ensure that the reagents are dry. Many forms of apparatus have been designed for macro-estimations, two of which are shown in figs. 12 and 13. The simple apparatus (fig. 12) is easy to operate and gives fairly accurate results (accuracy ± 5 %). A more elaborate design (fig. 13), which is a standard piece of apparatus in many laboratories, gives more accurate results (accuracy $\pm 1.5\%$), and a large number of determinations can be carried out rapidly. The reagent is made and stored in the apparatus, the inert atmosphere is introduced easily and the methane can be measured accurately.

The method has been adapted to the micro-estimation of the percentage of hydroxyl group.[1] The volume of methane which is measured is very small compared with the volume of the reaction vessel, and slight changes in temperature or pressure cause considerable movement in the mercury level. The reading of the mercury levels before and after the estimation must be at the same temperature, and before the readings are taken the reaction vessel is cooled to the temperature of a water-bath which is filled from a large supply kept at room temperature. The manometer is surrounded by a large water jacket. A burette, of 4–5 ml. capacity graduated in hundredths of a ml., is sealed on to a larger tube, of capacity 10–15 ml. This reservoir is necessary to take the increase in volume which occurs when the reaction vessel is heated to complete the reaction (fig. 14, p. 62).

1. Macro-determination

(a) Apparatus (fig. 12). The apparatus is dried and prepared for use as described below.

A blank determination is carried out under identical conditions as for the estimation.

A solution of the alcohol (to give approximately 50 ml. of methane) in pyridine, diisoamyl ether or anisole (concentration of the alcohol about 0.07 mol. per l.) is placed in A and the Grignard reagent (100% excess) in B. Pipettes are used for transferring these solutions. The upper surface of the ground-glass stopper is greased, the reaction vessel is closed and placed in a large beaker filled with water drawn from a store kept at room temperature. Nitrogen is passed through the apparatus

[1] J. Grant, *Quantitative Organic Micro-analysis based on the methods of Fritz Pregl*, 5th English ed., 1951, pp. 168–77; H. Roth, *Mikrochemie*, 1932, **5**, 140–56; A. Soltys, *ibid.* 1936, **20**, 107–25; W. Lüttgens and E. Negelein, *Biochem. Z.* 1934, **269**, 177–81; P. M. and G. F. Marrian, *loc. cit.* (p. 56, n. 3); B. Flaschenträger, *Z. physiol. Chem.* 1925, **146**, 219–26.

for 5 min., the tap *a* is then closed, *b* is kept open, and the mercury levels are read at half-minute intervals. When no further pressure change takes place it can be assumed that temperature equilibrium has been established. The height of

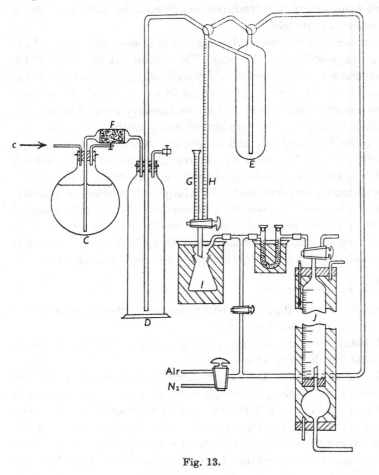

Fig. 13.

the mercury, the temperature of the water-bath, and the barometer reading are recorded. The water-bath is removed, the mercury reservoir is lowered slightly and the reaction started by tilting the flask to let the Grignard reagent flow into *A*. The mixture is shaken until no more gas is evolved. The pressure is kept as near to atmospheric as possible throughout the reaction by lowering the mercury reservoir gradually. The reaction vessel is placed in a water-bath at 70–100° C. for 10 min. and shaken frequently during heating. It is then allowed to cool, and finally placed in the original water-bath until it regains the initial temperature. When temperature equilibrium has been established again the mercury levels are adjusted to the same height and read.

Percentage of hydroxyl

$$= \frac{(\text{volume of methane in ml.}) \times 17 \times 100}{22 \cdot 4 \times 1000 \times \text{wt. of alcohol}} \times \frac{T_1 P_2}{T_2 P_1}.$$

(b) Apparatus (fig. 13)[1]. The Grignard reagent is prepared in flask C and transferred through D to the reservoir E by applying pressure at c. Excess magnesium is removed by passing through a filter of glass-wool, F. A solution of the alcohol is placed in the burette G, and the Grignard reagent is pumped from the reservoir into the burette H. Known volumes of the alcohol solution and the reagent are added to the reaction flask and the gas is collected in the burette J. The volume of methane formed is the change in volume minus the volume of the solutions added to the reaction flask.

2. *Micro-determination* (fig. 14)

See J. Grant, *Quantitative Organic Micro-analysis based on the methods of Fritz Pregl*, 5th English ed., 1951, pp. 168–77.

Accuracy: Macro., apparatus shown in fig. 12 ± 5 %.
Macro., apparatus shown in fig. 13 ± 1·5 %.
Micro., apparatus shown in fig. 14 ± 3 %.

PREPARATIONS: *Diisoamyl ether.* Commercial amyl ether dried over calcium chloride is boiled for several hours with sodium. The liquid is allowed to cool, and decanted from the sodium. Methyl magnesium iodide in amyl ether (5 ml.) is added and the mixture boiled again for several hours with fresh sodium. The ether is decanted from the sodium, mixed with phosphorus

[1] A detailed description of the use of the apparatus is given in *J. Amer. Chem. Soc.* 1930, **52**, 3736–8 and *ibid.* 1927, **49**, 3181–8 (E. P. Kohler and N. K. Richtmeyer; E. P. Kohler, J. F. Stone and R. C. Fuson).

pentoxide ($\frac{1}{10}$th the weight of ether), allowed to stand for several days, and finally fractionally distilled from sodium. B.p. 172–3° C.

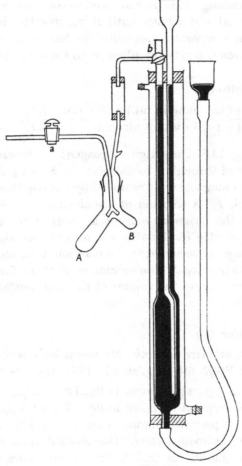

Fig. 14.

Pyridine. Pyridine, freed from its homologues by re-generating from the perchlorate,[1] is distilled under reduced pressure and shaken for 6 hr. with barium oxide (several pieces). A strong bottle fitted with a ground-glass stopper should be

---

[1] F. Arndt and P. Nachtwey, *Ber.* 1926, **59**, 448–55.

used. If solid barium oxide is left after shaking the pyridine is filtered rapidly through a fluted paper and stored for a few days over solid barium oxide. It is then ready for use. If no barium oxide remains after shaking more solid is added and the process is repeated.

Anisole. Anisole, dried over calcium chloride, is boiled for several hours with sodium and then fractionally distilled from sodium. B.p. 152–3° C. This is not sufficiently dry for use as a solvent in the above estimation, and it must be shaken immediately before use with phosphorus pentoxide ($\frac{1}{20}$th the weight of anisole). After allowing to stand the clear liquid is removed with a pipette and used.

The Grignard reagent.[1] Diisoamyl ether (100 g.) and magnesium turnings (12 g.) are placed in a three-necked flask (1 l.) fitted with a reflux condenser, a stirrer, an inlet tube for nitrogen and a dropping funnel. Nitrogen is passed through the apparatus and a solution of methyl iodide (60 g.) in diisoamyl ether (50 g.) is added slowly, with stirring, so that the temperature is kept below 20° C. If no reaction appears to take place after the addition of a little methyl iodide a few crystals of iodine are added and the reaction vessel warmed. After the final addition the mixture is heated on a vigorously boiling water-bath for 2 hr., allowed to cool to room temperature and diluted with diisoamyl ether (270 g.). The apparatus is converted for distillation by changing the condenser, and most of the excess methyl iodide is removed by distillation from a water-bath (about 30 min.). The reaction mixture is allowed to cool to room temperature and the clear liquid is filtered rapidly through glass-wool into a Claisen flask which has been dried at 110° C., allowed to cool in a vacuum desiccator and alternately evacuated and filled with nitrogen. The reagent can be pumped from the reaction flask through a filter of glass-wool by applying pressure (nitrogen), as it is in the apparatus shown in fig. 13 (*C* to *D*). The last traces of methyl iodide are removed by heating under reduced pressure for half an hour at 50° C. The Grignard reagent is allowed to cool, the flask is filled with nitrogen and the liquid

[1] Cf. E. P. Kohler and N. K. Richtmeyer, *J. Amer. Chem. Soc.* 1930, **52**, 3736–8.

decanted into a bottle (dried as above) filled with nitrogen. A tightly fitting rubber stopper is used to close the bottle which is kept in a vacuum desiccator.

The reagent is standardized by adding an excess of water in pyridine and measuring the volume of gas formed.

Nitrogen. Nitrogen from a cylinder is purified and dried before passing into the apparatus. For macro-determinations see E. P. Kohler, J. F. Stone and R. C. Fuson, *loc. cit.* (p. 61, n. 1); for micro-determinations see J. Grant, *loc. cit.* (p. 61).

Apparatus. The reaction vessel and pipettes are dried in an oven at 110° C. The reaction vessel is connected to the apparatus, the air is displaced by nitrogen and a slow current of gas is passed until the flask has cooled to room temperature.

A current of nitrogen is passed through the pipettes as they cool to room temperature.

B. Hydrogen from Lithium Aluminium Hydride

$$4R.OH + Li.Al.H_4 = 4H_2 + LiOR + Al(OR)_3.$$

Lithium aluminium hydride liberates hydrogen from compounds having an active hydrogen, and it can be analysed quantitatively by measuring the change of pressure caused on the addition of an excess of water.[1] Many simple alcohols, phenols, acids, mercaptans, primary and secondary amines and amino compounds also react quantitatively with the reagent and can be estimated in the apparatus shown in fig. 15 which measures the increase in pressure,[2] or in the usual Zerewitinoff apparatus.[3] The estimation is similar to that of Zerewitinoff and many advantages over the older method have been claimed, but a wider use is needed to show its value and how far it can replace other methods. The reaction appears to finish in a few seconds,

[1] J. A. Krynitsky, J. E. Johnson and H. W. Carhart, *Anal. Chem.* 1948, **20**, 311–12; cf. p. 55.

[2] J. A. Krynitsky, J. E. Johnson and H. W. Carhart, *J. Amer. Chem. Soc.* 1948, **70**, 486–9.

[3] See p. 58; H. E. Zaugg and B. W. Horrom, *Anal. Chem.* 1948, **20**, 1026–9; F. A. Hockstein, *J. Amer. Chem. Soc.* 1949, **71**, 305–7; micro-method, H. Lieb and W. Schöniger, *Mikrochemie*, 1950, **35**, 400–6; cf. *Anal. Chem.* 1948, **20**, 1022–6.

but it is necessary to reach equilibrium before and after the reaction and this may take up to an hour and a half. Subsequent estimations do not take as long and may be done in about half an hour. The results for the alcohols examined so far show good

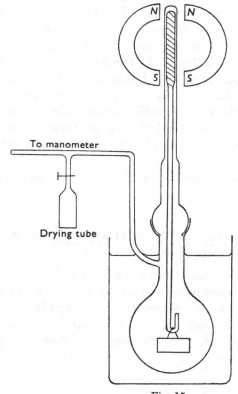

Fig. 15.

agreement with those found by other methods, including Zerewitinoff's. When keto-enol tautomerism is possible the method must be used with caution as the values for the active hydrogen do not agree with those found by bromine titration (see pp. 80–2) or by Zerewitinoff's method.

1. *By measuring the change in pressure*

A cold, diethyl ether solution of lithium aluminium hydride (about 100 ml., 0·3–0·7 M) is measured into the reaction flask

(500 ml.). The apparatus is assembled as shown in fig. 15. The flask and guide tube are clamped firmly to prevent leakage, and the flask is placed in a bath of crushed ice and water. The pinch clip is closed, and if any change in pressure occurs during 5 min. it is opened momentarily and the process repeated until equilibrium is established. This may take as long as 1 hr. The alcohol (about 0·1–0·4 mg., estimated to give an increase of pressure about 100 mm.) is weighed into the cup which is suspended from the supporting hook. The iron core is placed in the guide tube and held in the raised position until equilibrium is re-established (10–15 min.). The cup is then lowered and the alcohol brought into the solution of the reagent. The pressure is read at intervals ($\frac{1}{2}$–1 min.) until there is no further change. When the reaction takes place slowly the reaction flask can be shaken.

The volume of the apparatus is found in a preliminary experiment.

Active hydrogen

$$= \frac{\text{pressure change (in mm.)} \times \text{volume of apparatus (in ml.)}}{\text{millimoles of alcohol} \times 17,030}.$$

The constant 17,030 combines the values for R and T.

A second estimation can be carried out by hanging a new cup on another supporting hook and repeating the experiment.

PREPARATION: *Ethereal solution of lithium aluminium hydride.* See below and cf. A. E. Finholt, A. C. Bond, Jr., and N. I. Schlesinger, *J. Amer. Chem. Soc.* 1947, **69**, 1199–1203.

2. *By measuring the change in volume*

The active hydrogen can be determined by measuring the change in volume at constant pressure as described under Zerewitinoff's method (see pp. 55–64).

PREPARATION: *Ethereal solution of lithium aluminium hydride.*[1] A saturated solution is prepared by warming an excess of lithium aluminium hydride with di-*n*-butyl ether (150 ml.) which has been dried by distillation in dry nitrogen from lithium aluminium hydride.

[1] H. E. Zaugg and B. W. Horrom, *loc. cit.* (p. 64, n. 3).

COLORIMETRIC

A quantitative colorimetric method of estimation has been developed from the detection of alcohols by Agulhon's reagent.[1] A small percentage of an alcohol (0·1–1·0 %) can be estimated with a fair degree of accuracy, and the determination is unaffected by substances which normally interfere in other methods, even when they occur in large amounts. Aldehydes give a positive reaction with Agulhon's reagent, and the method cannot be used when this grouping is present, either in the free state, or as a polymer. Substances which are oxidized easily to aldehydes also interfere. Tertiary alcohols give a negative reaction and cannot be estimated. The following classes of compounds do not react although a reactive hydrogen is present: paraffin monocarboxylic acids, oxalic acid, citric acid, amino acids, amines and amino compounds, phenols, and aliphatic ketones.

The reagent is prepared by dissolving potassium dichromate in pure nitric acid.[2] The first stages in the reduction of the reagent are hardly noticeable, but at about 95 % reduction the colour changes rapidly and goes through green to blue, or violet-blue, which is used as the end-point.

A preliminary experiment to find an approximate end-point is carried out with a small volume (1 ml.) of the solution by adding increasing amounts of the reagent until a blue colour appears on standing for 5 min. A short series of samples are prepared with equal amounts of the solution (1 ml.) and varying amounts of the reagent (ranging on each side of the volume found in the above estimation). After leaving the mixtures to stand for 5 min. the tube is selected which just fails to reach the full blue colour. A comparison is made with a standard solution of the same alcohol.

Accuracy ± 5 %.

PREPARATION: *Agulhon's reagent*. Potassium dichromate (0·5 g.) is dissolved in pure nitric acid (100 ml., $d = 1·310$; freed

---

[1] E. C. Craven, *J. Soc. Chem. Ind.* 1933, **52**, 239–42T; cf. W. R. Fearon and D. M. Mitchell, *Analyst*, 1932, **57**, 372–4.

[2] H. Agulhon, *Bull. Soc. chim.* 1911 [v], **9**, 881–5 and *Ann. Chim. Anal.* 1912, **17**, 50.

from oxides of nitrogen by drawing a current of air through). The sensitivity of the reagent can be increased by adding concentrated nitric acid (3 vol.) to the reagent (1 vol.) (e.g. from 0·1 to 0·05 % for ethyl alcohol in water, and 0·3 to 0·1 % for methyl alcohol in water). An accurate ratio of the nitric acid ($d = 1·310$) to water is unnecessary, as dilution only makes the acid less susceptible to decomposition by sunlight.

MISCELLANEOUS

(1) A mixture of the alcohol and excess (at least 100 %) of phthalic anhydride in pyridine is heated under a reflux condenser at 100° C. for 1 hr., which is long enough to ensure complete reaction for most alcohols.[1] The mixture is left to cool and then titrated with a standard solution of sodium hydroxide using phenolphthalein as the indicator. Glycerol must be heated during 2 hr. for quantitative esterification, whilst a number of diols and tertiary alcohols dehydrate under the conditions of the reaction and cannot be estimated. The method is useful as the phenolic hydroxyl group does not react with phthalic anhydride, aldehydes interfere only slightly (cf. acetylation with acetic anhydride, p. 43) and good results are possible with fairly dilute solutions of alcohols.

(2) Alcohols can be estimated by measuring the volume of hydrogen liberated in the reaction with lithium aluminium hydride (see pp. 64–6). A method, using the same reagent, has been described recently in which the alcohol is added to an excess of the hydride which latter is back-titrated with a standard solution of an alcohol. Silver or platinum wires are used as indicator electrodes, and an isolated silver wire is used as a reference electrode. Tetrahydrofuran is a suitable solvent and the end-point is indicated by a sharp change in the reduction potential.[2] If oxygen is not excluded completely from the titration flask some of the hydride is lost by oxidation.[3] The results show fair agreement with theoretical values, being between 1 and 5 % high.

[1] P. J. Elving and B. Warshowsky, *Anal. Chem.* 1947, **19**, 1006–10.
[2] C. J. Lintner, R. H. Schleif and T. Higuchi, *Anal. Chem.* 1950, **22**, 534–8.
[3] T. Higuchi, *ibid.* 1950, **22**, 955.

(3) The red co-ordination complex which is formed when an alcohol is added to a solution of ammonium hexanitratocerate has been used as a qualitative test for alcohols.[1] A colorimetric method of estimation has been developed from this test and used for a few of the lower aliphatic alcohols.[2] The low stability of the colour and interference from many of the commoner reducing agents are serious disadvantages.

(4) A mixture of the alcohol and acetyl chloride in toluene is heated under a reflux condenser for 2 hr. The reaction mixture is left to cool and then titrated with a standard solution of alkali.[3] The apparatus is of a special design to exclude moisture, and has a trap containing a measured volume of a standard solution of sodium hydroxide to prevent the loss of hydrogen chloride or acetic acid.

(5) A mixture of the alcohol, an excess of benzoyl chloride and tetralin, is heated under a reflux condenser connected to wash bottles containing a known volume of a standard solution of alkali. A dry, inert gas (nitrogen or hydrogen) is passed slowly through the reaction mixture until no more hydrogen chloride is formed. The acid, which is a measure of the percentage of hydroxyl, is absorbed in the standard alkali and the excess is determined from a titration with standard acid.[4]

Accuracy $\pm 10\%$.

(6) The sample (2–5 mg.) is added to a solution of lithium aluminium hydride in n-propyl alcohol and the hydrogen formed is carried by nitrogen into a combustion tube filled with cupric oxide heated to 1100–1200° C. The equivalent of water which is formed is passed over red-hot coke and converted into carbon monoxide which reacts with iodine pentoxide to liberate iodine which is estimated by titration with sodium thiosulphate.[5]

[1] F. R. Duke and G. F. Smith, *Ind. Eng. Chem.* (Anal. ed.), 1940, **12**, 201–3.
[2] F. R. Duke, *ibid.* 1945, **17**, 572–3.
[3] B. L. Johnson, *ibid.* 1948, **20**, 777–8; cf. H. Burkett, *Proc. Indiana Acad. Sci.* 1949, **58**, 142.
[4] T. M. Meijer, *Rec. trav. chim.* 1934, **53**, 387–97.
[5] W. Schöniger, *Z. anal. Chem.* 1951, **133**, 4–7.

Specific for Methyl Alcohol

I. Many methods have been described for the determination of small amounts of methyl alcohol; however, oxidation and estimation by Schiff's reagent of the formaldehyde so formed is the most sensitive. This method was proposed originally in 1910,[1] and has been modified since and improved by many investigators.[2] The estimation was developed originally, and modified since, for the determination of methyl alcohol in fermented liquors, extracts and medicinal tinctures, all of which contain large amounts of ethyl alcohol. Acetaldehyde also gives a colour with Schiff's reagent and so interferes with the analysis. The conditions can be adjusted to stop the colour developing from the relatively large amount of acetaldehyde, but then the sensitivity decreases considerably. In finding the concentration of methyl alcohol vapour in air this difficulty usually does not arise.

The solution of methyl alcohol is oxidized with an excess of acid potassium permanganate, but in dilute solution these conditions yield practically no formaldehyde unless a little ethyl alcohol (0·005 ml.) is added. The two alcohols are oxidized together and the production of formaldehyde is increased to a measurable value. The excess of the oxidizing agent is reduced with oxalic acid, the hydrogen-ion concentration is adjusted by the addition of sulphuric acid, and the colour is developed with Schiff's reagent. The conditions given below are the optimum to produce the maximum intensity from formaldehyde and at the same time produce no colour from the oxidation of the ethyl alcohol. The oxidation of the two alcohols was carried out originally in the presence of sulphuric acid, but it was shown

[1] G. Denigès, *Compt. rend.* 1910, **150**, 832–4.

[2] C. Simmonds, *Analyst*, 1912, **37**, 16–18; G. C. Jones, *ibid.* 1915, **40**, 218–21; C. M. Jephcott, *ibid.* 1935, **60**, 588–92; E. Elvove, *J. Ind. Eng. Chem.* 1917, **9**, 295–7; R. M. Chapin, *ibid.* 1921, **13**, 543–5; F. R. Georgia and R. Morales, *ibid.* 1926, **18**, 304–6; L. O. Wright, *ibid.* 1927, **19**, 750–2; W. L. O. Whalley, *ibid.* 1928, **20**, 320–2; A. Kling and A. Lassieur, *Compt. rend.* 1924, **178**, 1006–9; D. L. Smith and B. F. Banting, *J. Amer. Chem. Soc.* 1929, **51**, 129–39; E. M. Iofinova-Gol'dfein, *Hig. Truda i Tekh. Bezopasnosti*, 1936, **14**, no. 1, 82–4; and *Chemie & industrie*, 1936, **37**, 658; T. von Fellenberg, *Congr. intern. tech. chim. ind. agr.*, *Compt. rend. Ve Congr.* 1937, **1**, 184–96; S. L. Ginzburg, *Org. Chem. Ind.* (U.S.S.R.), 1939, **6**, 177–9.

later that a lower hydrogen-ion concentration increases the ratio of formaldehyde to acetaldehyde, and the best results are obtained with a minimum concentration of phosphoric acid.

A comparison of the colours produced by standard solutions of methyl alcohol and formaldehyde shows that the oxidation of the methyl alcohol to formaldehyde is not quantitative and that the depth of colour is not proportional to the concentration. It is important therefore to make up a number of standards covering the concentration of the unknown and to prepare all the solutions at the same time under identical conditions.

Aldehydes, *iso*butyl, amyl and propyl alcohols, acetic and tartaric acids do not interfere in the concentrations found normally in fermented liquors. A large number of essential oils yield formaldehyde on oxidation and contribute to the final colour.

Methyl alcohol (10 ml., at least 30 p.p.m.), dilute ethyl alcohol (1 ml., 5 % by volume), potassium permanganate (5 ml., 1 %) and dilute phosphoric acid (1 ml., 25 ml. of 85 % acid made up to 100 ml. with water) are mixed in a graduated flask (25 ml.). The oxidation takes place at room temperature during 1 hr. if the flask is shaken occasionally. A solution of oxalic acid (1 ml., 5 %) is added, the flask is shaken, and when the solution becomes colourless, or nearly so, dilute sulphuric acid (2 ml., 30 ml. of concentrated acid made up to 100 ml. with water) and Schiff's reagent (5 ml.) are added. The solution is mixed thoroughly, and after standing for 3 hr. it is compared in a colorimeter with a standard solution of approximately the same concentration.

Methyl alcohol vapour can be removed quantitatively from air by bubbling through water. When the final concentration is too low for estimation it can be increased by distillation, but if possible a larger volume of air should be used.

PREPARATION: *Schiff's reagent.* Fuchsin (0·5 g.) is dissolved in warm water (400 ml.), and after allowing to cool sodium bisulphite (2 g., anhydrous) is added and the liquid is stirred until a solution is obtained. Concentrated sulphuric acid (4 ml.) is added, with stirring, and the solution is left overnight in a securely stoppered, brown-glass bottle. A clear solution is formed and the reagent is ready for use.

II. Small amounts of methyl alcohol can be estimated in the presence of very large amounts of ethyl and higher aliphatic alcohols by forming the corresponding iodides from the primary alcohols. The alkyl iodides are distilled, silver acetate is added to the distillate and the silver iodide filtered off and weighed. The alcohols are oxidized in the cold and the oxygen equivalent determined. The estimation is long and tedious but delicate.[1]

III. Methyl alcohol can be estimated by adding an excess of sodium hypobromite and determining the excess by an iodometric titration.[2]

IV. Methyl alcohol reacts rapidly with nitrous acid to give a quantitative yield of methyl nitrite which latter can be extracted and estimated. The nitrite is decomposed with acid, and the liberated nitrous acid is determined either iodometrically by adding potassium iodide and titrating the free iodine with sodium thiosulphate or colorimetrically by forming an azo dye with sulphamic acid and α-naphthylamine.[3]

ENOLS

The equilibrium which exists between keto and enol forms increases the difficulty of analysis by chemical means. The conversion and removal of one form destroys the equilibrium and leads to the continuous change of the other form either until equilibrium is re-established or until there is no further reaction. The preparation and analysis of the copper co-ordination complex, bromination and any method which removes the enol form must be done rapidly for an accurate estimate of the composition of the tautomeric mixture. It is not always possible to get reliable results by bromination, even when the conditions

[1] M. Flanz, Ann. fals. 1935, 28, 260–77.
[2] K. Száhlender, Magyar Gyógyszerésztud. Társaság Értesitöje, 1933, 9, 125–47.
[3] W. Ender, Angew. Chem. 1934, 47, 227–8; V. K. Nizovkin and O. I. Okhrimenko, Lesokhim. Prom. 1939, 2, no. 7, 17; I. Ya. Shaferstein, J. Appl. Chem. (U.S.S.R.), 1934, 7, 239–50.

used normally give good results, as when the rate of change of keto to enol is very rapid (e.g. ethyl oxalacetate). Estimation by the physical method of measuring the molecular refractivity of the mixture does not change the composition of the mixture. It is used frequently and does not require the same skill as the chemical methods to give accurate and reproducible results.

Enols can be quantitatively analysed by the following methods:

Determination of the molecular refractivity.

Preparation and estimation of a copper co-ordination complex:

$$2R.CO.CH_2.COOEt + Cu(OAc)_2 = \begin{array}{c} OEt \\ | \\ C=O \\ CH \\ C-O \\ | \\ R \end{array} Cu \begin{array}{c} R \\ | \\ O-C \\ CH + 2HOAc. \\ O=C \\ | \\ OEt \end{array}$$

Bromination:

$$R.\underset{\underset{OH}{|}}{C}=CH.COOEt + Br_2 = \left[R.\underset{\underset{OH}{|}}{\overset{\overset{Br}{|}}{C}}-\underset{\underset{H}{|}}{\overset{\overset{Br}{|}}{C}}-COOEt \right] \longrightarrow R.CO.CHBr.COOEt.$$

$R.CO.CHBr.COOEt + 2HI = HBr + I_2 + R.CO.CH_2.COOEt.$

Colorimetric.

Miscellaneous.

DETERMINATION OF THE MOLECULAR REFRACTIVITY

The percentage of enol in a tautomeric mixture can be found by comparing the theoretical values of the molecular refractivity of the keto and enolic forms with the observed value. The theoretical values for the desmotropes are calculated as the sum of the constituent refractivities associated with the atoms and bonds in the structures.[1]

With compounds forming a conjugated system in the enolic form (e.g. 1:3-diketones and β-ketonic esters) values for the percentage of enol are obtained which apparently do not agree

[1] F. Eisenlohr, Z. physikal. Chem. 1910, 75, 585–607; W. Swietoslawski, J. Amer. Chem. Soc. 1920, 42, 1945–51.

with those found by other methods. This optical anomaly arises from the conjugated double bonds which introduce another additive factor, and since the observed value is generally higher than that calculated from the known equivalents the compounds are said to exhibit optical exaltation.[1] When the optical exaltation is taken into account the results agree with those obtained by other methods.

The molecular refraction is calculated by substituting the observed refractive index and density in either of the equations:

$$R = \frac{M}{d}\,(n-1),$$

$$R = \frac{M}{d}\frac{n^2-1}{n^2+2},$$

where n = refractive index and d = density at the same temperature. The first[2] was derived experimentally, and has been replaced completely by the second which was deduced from the electromagnetic and wave theories of light.[3] The values of R from the two equations are different, and the following table refers only to those calculated from the Lorentz-Lorenz equation which show less variation with change of state.

The refractive index is determined accurately in either the Abbé or Pulfrich refractometer (see pp. 75–8), and the density is measured using a pyknometer (see p. 78).

Percentage of enol

$$= 100\ \frac{\text{observed value} - \text{theoretical keto value}}{\text{theoretical enol value} - \text{theoretical keto value}}.$$

The atomic refractivities given in the table on p. 75 are for the sodium (D) line, and for the red and blue lines of the hydrogen spectrum (H_α, H_β, H_γ).

[1] K. v. Auwers, *Ber.* 1911, **44**, 3514–24; K. v. Auwers, *Ann.* 1918, **415**, 169–232; K. v. Auwers and H. Jacobson, *ibid.* 1921, **426**, 161–236; cf. H. W. Post and G. A. Michalek, *J. Amer. Chem. Soc.* 1930, **52**, 4358–62; K. v. Auwers, *ibid.* 1931, **53**, 1496–1500.
[2] J. H. Gladstone and T. P. Dale, *Phil. Trans.* 1858, **148**, 887–94; 1863, **153**, 317–43; H. Landolt, in 1864.
[3] H. A. Lorentz, in 1880; L. V. Lorenz, in 1880.

	D	H_α	H_β	H_γ
Carbon	2·418	2·413	2·438	2·466
Hydrogen	1·100	1·092	1·115	1·122
Oxygen (in CO group)	2·211	2·189	2·247	2·267
Oxygen (in ethers)	1·643	1·639	1·649	1·662
Oxygen (in OH group)	1·525	1·522	1·531	1·541
Nitrogen (prim. amino)	2·322	2·309	2·368	—
Nitrogen (sec. amino)	2·502	2·475	2·561	—
Nitrogen (tert. amino)	2·840	2·807	2·940	—
Nitrogen (nitrile)	3·118	—	—	—
Chlorine	5·967	5·933	6·043	6·101
Bromine	8·865	8·803	8·999	9·152
Iodine	13·900	13·757	14·224	14·521
Double bond	1·733	1·686	1·824	1·893
Triple bond	2·398	2·328	2·506	2·538

Optical exaltation as observed in β-ethoxycrotonic ester, 1·8.

1. The Abbé refractometer

The liquid is placed in contact with the hypotenuse of the right-angled prism ABC (fig. 16). Light enters the prism at grazing incidence, and on emerging is observed by means of a fixed telescope. When $n = N \sin A$ the light will emerge normally from the face AB, but with a change in the value of the refractive index (n to N) the light will be at angle ϕ to AB, and will only be observed in the telescope by rotating the prism through an angle θ. By measuring this angle of rotation the refractive index of the liquid (n) is calculated:

$$\hat{A} = \alpha + \phi,$$

$$N = \frac{\sin \theta}{\sin \phi}, \quad \therefore \frac{\sin \theta \cos \phi}{\sin \phi} = \sqrt{(N^2 - \sin^2 \theta)}.$$

$$n = N \sin \alpha$$

$$= \frac{\sin \theta}{\sin \phi} \sin (A - \phi)$$

$$= \frac{\sin \theta \sin A \cos \phi - \sin \theta \sin \phi \cos A}{\sin \phi}$$

$$= \sin A \sqrt{(N^2 - \sin^2 \theta)} - \cos A \sin \theta.$$

The instrument is usually calibrated so that the refractive index can be read directly from the scale.

The instrument is opened, and arranged so that the surface of the fixed prism is horizontal; a drop of the liquid is placed on this surface and the movable prism clamped into position. The heating circuit is switched on, and when the desired temperature has been reached and has remained constant for some time, the prism is illuminated and rotated until the line of separation between light and dark is exactly on the centre of the cross-wires. The refractive index is read on the scale, the fourth decimal place being estimated.

Micro-determinations can be carried out with this apparatus, as only one drop of liquid is needed.

Accuracy ± 0.0001.

2. *The Pulfrich refractometer*

The liquid is placed in a cell cemented to the top of a glass prism through which light enters at grazing incidence and is observed by a movable telescope (fig. 17). By measuring the angle of emergence the refractive index is calculated as follows:

$$\frac{\sin i}{\sin r} = \frac{N}{n},$$

$$\frac{\sin 90°}{\sin r} = \frac{N}{n},$$

$$\therefore \ n = N \sin r. \tag{1}$$

$$\sin r = \cos \phi, \quad N = \frac{\sin \theta}{\sin \phi}.$$

$$\therefore \ \frac{\sin \theta \cos \phi}{\sin \phi} = \sqrt{(N^2 - \sin^2 \theta)}.$$

Substituting for N and $\sin r$ in (1),

$$n = \frac{\sin \theta}{\sin \phi} \cos \phi = \sqrt{(N^2 - \sin^2 \theta)}.$$

In the determination of the refractive index monochromatic light must be used, as the degree of refraction depends upon the wave-length.

Two readings are necessary to determine the angle of emergence, one to correct for any error in the zero point of the instrument, and the second to observe the experimental value. The zero correction is done with the help of a small right-angled prism let into the side of the movable telescope. The disk is rotated so that its zero coincides with that of the vernier; light from a small lamp or torch is admitted through the prism, and the field of view (fig. 18) adjusted so that the upper cross-wire

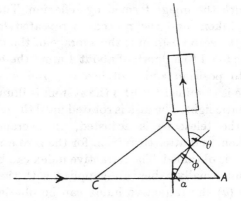

n=Refractive index of the liquid
N=Refractive index of the prism

Fig. 16.

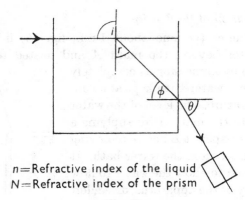

n=Refractive index of the liquid
N=Refractive index of the prism

Fig. 17.

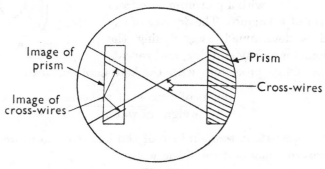

Image of prism

Prism

Cross-wires

Image of cross-wires

Fig. 18.

coincides with the image formed by reflexion. The reading on the disk is taken, and the procedure repeated for the lower cross-wire; the zero position is the average of the two readings.

The cell is filled to a depth of about 4 mm., the heating cell is lowered into position and switched on, and when a steady temperature is obtained ($\pm 0\cdot1°$) the system is illuminated with monochromatic light. The disk is rotated until the refracted light is visible, the telescope is adjusted, an average of several readings taken, and after correction for the zero error the angle of emergence is obtained. The refractive index can be calculated as above, but usually tables are supplied with the instrument from which (a) the refractive index can be obtained knowing the angle of emergence, and (b) a correction for temperature variation can be calculated.

Accuracy $\pm 0\cdot00002$.

3. Determination of the density

A pyknometer (volume about 1 ml., fig. 19) is filled with distilled water beyond the mark A and heated to a known

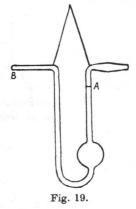

temperature by immersing as completely as possible in a water-bath kept at a constant temperature. The level of the water is adjusted to the mark A by applying a piece of filter paper at B before removing the pyknometer from the water-bath. It is then inclined to allow the water to flow into the body of the apparatus, dried, left to cool to room temperature, and weighed by suspending with a platinum wire from the arm of a balance. The density of the liquid is determined by weighing the pyknometer empty and dry, and repeat-

Fig. 19.

ing the above procedure with the keto-enol mixture:

$$d_y^x = \frac{\text{weight of keto-enol mixture}}{\text{weight of water}},$$

where $x =$ constant temperature of the keto-enol mixture, and $y =$ constant temperature of the water.

PREPARATION AND ESTIMATION OF A COPPER CO-ORDINATION COMPLEX

$$2R.CO.CH_2.COOEt + Cu(OAc)_2 = \begin{array}{c} OEt \\ | \\ C=O \\ CH \\ \diagdown \\ C-O \\ | \\ R \end{array} \quad Cu \quad \begin{array}{c} R \\ | \\ O-C \\ CH \\ \diagup \\ O=C \\ | \\ OEt \end{array} + 2HOAc$$

The enol in an equilibrium mixture of the keto and enol forms can be determined by preparing its copper co-ordination complex which is soluble in non-polar but insoluble in polar solvents.[1] A mixture of an aqueous-alcoholic solution of copper acetate and chloroform is added to a well-cooled alcoholic solution of the tautomeric mixture. A co-ordination compound is formed in the aqueous layer and is extracted with the chloroform, whilst the keto form does not react and remains in the aqueous alcohol. The two layers are run off giving a complete separation of the two components. The percentage of enol is found from the amount of copper in the chloroform extract. The formation of the copper co-ordination compound stops immediately the chloroform and aqueous solutions separate into two layers, as there is no extraction of copper acetate from the aqueous-alcoholic layer and there is no measurable reaction between the enol regenerated subsequently in the chloroform layer and the copper acetate in the aqueous layer.

The accuracy of the above method depends largely on the skill of the operator and the time taken for the formation of the co-ordination compound and the separation of the two immiscible liquids.

A mixture of an aqueous solution of copper acetate (6–10 ml.), alcohol (12 ml.) and chloroform (6–8 ml.) is cooled to $-10°$ C. and then added as rapidly as possible to an alcoholic solution of the tautomeric compound. The mixture is shaken, poured into a separating funnel containing water (250 ml.) (this should be carried out as rapidly as possible, 5–10 sec.), shaken a few more times, and the chloroform layer run off. An exact separation

[1] W. Hieber, *Ber.* 1921, **54**, 902–12.

can be effected by rotating the funnel, and the last traces of the co-ordination compound can be collected by rinsing the funnel with a little chloroform. The percentage of enol is found by estimating the cupric ion formed by decomposing the co-ordination compound with dilute sulphuric acid (10 %). Potassium iodide (2–3 g.) is added to the acid solution, and the free iodine is titrated with a standard solution of sodium thiosulphate using starch as the indicator (cf. p. 120).

The whole estimation should be completed in 5–10 min.

Accuracy ± 5 % (see above).

PREPARATION: *Copper acetate solution.* Copper acetate (50 g., $Cu(OAc)_2$, $1H_2O$) is dissolved in water and the solution made up to 1 litre. The solution should be standardized roughly before each estimation, and about twice the amount needed to combine with the enol should be used.

BROMINATION

$$R.C{=}CH{-}COOEt + Br_2 = \left[\begin{array}{c} \underset{\overset{|}{OH}}{\overset{Br}{\underset{}{C}}}{-}\underset{\overset{|}{H}}{\overset{Br}{\underset{}{C}}}{-}COOEt \end{array} \right] \longrightarrow R.CO.CHBr.COOEt$$

$R.CO.CHBr.COOEt + 2HI = HBr + I_2 + R.CO.CH_2.COOEt.$

The percentage of an enol in a tautomeric mixture can be estimated approximately by a direct titration with bromine.[1] A standard alcoholic solution of bromine is added to the keto-enol mixture dissolved in the same solvent until a faint yellow colour persists at the end-point. The amount of enol so determined is always high owing to the equilibrium being destroyed on the addition of bromine and the continuous formation of enol throughout the titration. The error is variable depending upon the particular keto-enol system, but it can be reduced considerably by carrying out the titration at a low temperature, and so reducing the velocity of the tautomeric change. The solution of bromine must be standardized immediately before the titration owing to the reaction between alcohol and bromine.

---

[1] K. H. Meyer, *Ann.* 1911, **380**, 212–42; K. v. Auwers and F. Eisenlohr, *Z. physiol. Chem.* 1913, **83**, 429–41.

The accuracy can be improved greatly by using a slightly different procedure. Excess of an alcoholic solution of bromine is added rapidly to a cooled solution of the tautomeric mixture, the excess is removed immediately (see below) and potassium iodide and concentrated hydrochloric acid are added.[1] A bromo-ketone having the grouping —CO—C(Br.R)—CO.R'— is formed from the enol in the equilibrium mixture in the initial reaction, and is estimated subsequently by the liberation of an equivalent of iodine. The reduction is carried out easily and quantitatively with hydrogen iodide, and the iodine is estimated by titration with a standard solution of sodium thiosulphate. The accuracy of the method depends largely on the speed with which the excess of bromine is removed, as more of the enol is formed from the keto desmotrope as soon as the equilibrium is disturbed. A number of reagents, including sodium thio-sulphate,[2] an alcoholic solution of β-naphthol[3] and di*iso*butyl-ene,[4] have been used to combine with and remove the excess of bromine. Alcoholic β-naphthol is more satisfactory than sodium thiosulphate, but even when pure β-naphthol is used the alcoholic solution turns brown rapidly, and often difficulty is experienced in seeing when all the bromine has been removed and in observing the end-point in the final titration. It is claimed that greater accuracy is possible with di*iso*butylene as the olefine does not form a stable compound with iodine, and as the dibromide is colourless no difficulty arises from coloured solutions.[5]

The method cannot be used for enols which do not react, or only react slowly, with bromine (e.g. some esters of diaceto-succinic acid).[6]

1. *Direct titration with bromine*

An alcoholic solution of the enol is cooled to about $-10°$ C. and titrated with a freshly prepared solution of bromine in

[1] K. H. Meyer, *loc. cit.* (p. 80, n. 1); K. H. Meyer and R. Kappelmeier, *Ber.* 1911, **44**, 2718–24.
[2] K. H. Meyer, *loc. cit.* (p. 80, n. 1).
[3] K. H. Meyer and P. Kappelmeier, *loc. cit.* (p. 81, n. 1).
[4] S. R. Cooper and R. P. Barnes, *Ind. Eng. Chem.* (Anal. ed.), 1938, **10**, 379.
[5] *Ibid.*
[6] H. P. Kaufmann and E. Richter, *Ber.* 1925, **58**, 216–22.

alcohol (0·2N) until a faint yellow colour persists. The alcoholic solution of bromine is unstable and must be standardized immediately before the titration. Solutions made up with methyl alcohol are more stable than those prepared with ethyl alcohol.

2. Indirect titration with bromine

A solution of the enol in absolute methyl alcohol is cooled to −10° C., and a methyl alcoholic solution of bromine (0·1N approx.) is added quickly until a faint yellow colour persists. An alcoholic solution of β-naphthol (2 ml., 10 %) or diisobutylene (slight excess) is added immediately so that the time taken to add the bromine and absorb the excess is about 15 sec. Potassium iodide (10 ml., 5 %) is added and the mixture is warmed to 30° C. on a water-bath for 20 min. (preferably in a dark room). The free iodine is estimated with a standard solution of sodium thiosulphate (0·1N) using starch as the indicator.

Accuracy 7 % (β-naphthol), 4–5 % (diisobutylene).

COLORIMETRIC

The enolic forms of β-ketonic esters and 1:3-diketones react in solution with metallic salts giving highly coloured compounds which can be used for their colorimetric estimation.[1] An alcoholic solution of ferric chloride is used generally, and the maximum intensity of colour is developed when the reaction takes place between equimolecular proportions of the reagent and the enol. The standard solutions by which comparison is made are of three types:

(a) A solution prepared from sublimed ferric chloride and the pure enol.

(b) A solution of a ferric salt of the type FeR_3 (where R is an enolic residue) to which 2 mol. of ferric chloride and 3 mol. of hydrochloric acid are added:

$$FeR_3 + 2FeCl_3 = 3FeRCl_2.$$

[1] W. Wislicenus, Ber. 1899, 32, 2837–40; cf. L. Knorr and H. Schubert, Ber. 1911, 44, 2772–8; L. Knorr and H. P. Kaufmann, Ber. 1922, 55, 232–48.

(c) A solution of the tautomeric equilibrium mixture, usually in alcohol, to which 1 mol. of ferric chloride is added. The percentage of enol is determined by comparison with standards (a) or (b).

The standard solution (c) is stable usually for several months. Solutions (a) and (b) may deteriorate although not seriously unless decomposition occurs (e.g. dibenzoylacetylmethane).

The method can be applied only to those enols which react with ferric chloride. If the reaction takes place slowly it can be speeded up usually by heating to about 30° C., but enols from the esters of diacetosuccinic acid do not react with ferric chloride even under these conditions. They can be analysed quantitatively by the bromine titration methods.[1]

An alcoholic solution of ferric chloride ($0 \cdot 1$ N) is added to the keto-enol mixture dissolved in the same solvent until the maximum intensity of colour develops. When the complex forms slowly the solution can be heated to about 30° C. before the alcoholic ferric chloride is added. The estimation is carried out colorimetrically by comparison with standard solutions (see above p. 67).

Accuracy $\pm 5 \%$.

MISCELLANEOUS

(1) Free thiocyanogen, $(CNS)_2$, has been used for the estimation of enols.[2] The method is inferior to analysis by bromine titration.

(2) The constitution of keto-enol mixture has been investigated by ozonolysis at $-20°$ C., and the method has been used for the semi-quantitative estimation of the percentage of enol present in an equilibrium mixture.[3]

(3) The keto-enol mixture and an acidified solution of potassium bromide-potassium bromate are mixed and diluted simultaneously in a mixing chamber and then passed over

[1] See above, p. 80; H. P. Kaufmann and E. Richter, Ber. 1925, 58, 216–22.
[2] H. P. Kaufmann and G. Wolff, Ber. 1924, 57, 934–7.
[3] J. Scheiber and P. Harold, Ber. 1913, 46, 1105–10; Ann. 1914, 405, 295–346.

a platinum electrode. The relative amounts of the two solutions are adjusted so that the potential, as measured by the electrode, rises sharply; this corresponds to the end-point of the titration of the enol by bromine.[1] The method is reported to give good results for mixtures even when the enol content is as low as 10^{-5} %.

PHENOLS

Substitution reactions of the aromatic nucleus as well as the reactions of the hydroxyl group are used for the estimation of phenols. Bromination is rapid and quantitative in the *ortho*- and *para*-positions and can be done conveniently with an acidified bromide-bromate solution. Esterification with acetic anhydride or acetyl chloride and measuring the volume of gas formed in the reaction with either methyl magnesium iodide or lithium aluminium hydride are methods similar to those used for alcohols. The coloured products of the coupling reactions give many sensitive methods of analysis. Gibbs' method of preparing and comparing the intensely blue solutions of the indophenols can be used for concentrations as low as five parts in a billion. The following general methods are used for the estimation of phenols:

Esterification:

A. Acetylation with acetic anhydride:
$$R.OH + (CH_3CO)_2O = CH_3.COOR + CH_3.COOH.$$

B. Acetylation with acetyl chloride:
$$R.OH + CH_3.COCl = CH_3.COOR + HCl.$$

Measurement of gaseous reaction products:

A. Methane from methyl magnesium iodide—Zerewitinoff's method:
$$R.OH + CH_3.Mg.I = CH_4 + Mg{\overset{\diagup I}{\diagdown OR}}.$$

B. Hydrogen from lithium aluminium hydride:
$$4R.OH + Li.Al.H_4 = 4H_2 + LiOR + Al(OR)_3.$$

[1] G. Schwarzenbach and Ch. Wittwer, *Helv. Chim. Acta*, 1947, **30**, 656–8.

Bromination:

$$\text{(phenol)} + 4Br_2 = \text{(tribromophenol bromide, OBr, Br, Br, Br)} + 4HBr,$$

$$\text{(OBr, Br, Br, Br)} + HI = \text{(OH, Br, Br, Br)} + I + Br.$$

Colorimetric:

A. Gibbs' method:

$$O={<}{=}N.Cl + {<}OH = O={<}{=}N{-}{<}{-}OH + HCl.$$

B. Coupling with diazo compounds:

$$C_6H_5.OH + Cl.N_2R' = R'{-}N{=}N{-}{<}{-}OH.$$

C. With Millon's reagent.

D. Formation of nitrosophenols with nitrous acid.

E. With the Folin-Denis reagent.

Miscellaneous.

ESTERIFICATION

A. Acetylation with Acetic Anhydride

$$R.OH + (CH_3CO)_2O = CH_3.COOR + CH_3.COOH,$$

B. Acetylation with Acetyl Chloride

$$R.OH + CH_3COCl = CH_3.COOR + HCl.$$

Phenols can be estimated by methods which involve acetylation as described under alcohols (see pp. 42–55). 2:4:6-Trisubstituted phenols usually react slowly and incompletely with acetic anhydride and should not be estimated by this method.

MEASUREMENT OF GASEOUS REACTION PRODUCTS

A. Methane from Methyl Magnesium Iodide: Zerewitinoff's Method

$$R.OH + CH_3.Mg.I = CH_4 + Mg\diagup^{I}_{\diagdown OR}.$$

The method of Zerewitinoff can be used for the estimation of phenols (see pp. 55–64). Diamyl ether should not be used as the solvent when the reaction product is precipitated, as some unreacted phenol usually is occluded in the solid. Anisole, pyridine or a mixture of anisole and pyridine (5:1) are usually satisfactory. A blank determination must be carried out to make sure that the solvent is dry.

B. Hydrogen from Lithium Aluminium Hydride

$$4R.OH + Li.Al.H_4 = 4H_2 + LiOR + Al(OR)_3.$$

Lithium aluminium hydride can be used for the estimation of phenols as described under alcohols (see pp. 64–6). Variable results, as with hydroquinone, may be caused by the formation of insoluble compounds which cover the phenol and prevent further reaction. Resorcinol reacts incompletely and gives only half the expected value, possibly owing to the above cause or from partial reaction in the keto form.[1]

BROMINATION

The introduction of a hydroxyl group into the benzene ring makes possible the rapid and quantitative replacement of the

[1] Cf. Th. Zerewitinoff, *Ber.* 1908, **41**, 2233–43.

ortho- and *para*-nuclear hydrogens with bromine. The reaction was used for the first time for the estimation of phenols in 1876.[1] Two methods of estimation are described below. In the first an acidified solution of the phenol is titrated with a standard potassium bromide-potassium bromate solution. In the second an excess of the bromide-bromate solution is added to an alkaline solution of the phenol, the mixture is acidified and after the reaction has finished the excess of bromine is determined. The direct titration method is useful occasionally when the more normal back-titration cannot be used (see p. 89). The slowness of the reaction near the end-point, the necessity for a small excess of bromine to persist for 2–4 min. and the use of an external indicator all combine to make the procedure tedious. The values are often high from loss of bromine, and for these reasons the indirect method is used whenever possible. The reaction has been the subject of numerous investigations and correct values are found for many phenols, but some results show that there is much about the reaction that is not clear.[2]

Anomalous results are found when certain groups are substituted in the *ortho*- and *para*-positions. Salicylic acid and *o*- and *p*-hydroxybenzaldehyde brominate with the breaking off of the carboxyl and formyl groups and the formation of tribromophenol.[3] It appears that the displacement of groupings, such as carboxylic acid, sulphonic acid, aldehyde and hydroxymethylene, can either be made quantitative or prevented, depending on the experimental conditions. The use of a powerful oxidizing agent, such as bromine, can therefore cause undesirable side reactions. These can frequently be reduced to a minimum by cooling the reaction mixture in an ice-bath, but low values may

[1] W. F. Koppeschaar, *Z. anal. Chem.* 1876, **15**, 233–45.
[2] A. R. Day and W. T. Taggart, *J. Ind. Eng. Chem.* 1928, **20**, 545–7; L. V. Redman, A. J. Weith and F. P. Brock, *ibid.* 1913, 5, 389–93; C. M. Pence, *ibid.* 1912, **4**, 518–20; I. W. Ruderman, *Ind. Eng. Chem.* (Anal. ed.), 1946, **18**, 753–9; R. D. Scott, *ibid.* 1931, 3, 67–70; M. M. Sprung, *ibid.* 1941, **13**, 35–8; A. W. Francis and A. J. Hill, *J. Amer. Chem. Soc.* 1924, **46**, 2498–505; J. J. Fox and H. F. Baker, *J. Soc. Chem. Ind.* 1918, **37**, 268–72T; T. Callan and J. A. R. Henderson, *ibid.* 1922, **41**, 161–4T; M. Beukema-Goudsmit, *Pharm. Weekblad*, 1934, **71**, 380–91.
[3] A. H. Allen, *Commercial Organic Analysis*, 4th ed. 3, 480–2; A. W. Francis and A. J. Hill, *loc. cit.* (p. 87, n. 2).

result owing to the velocity of reaction between bromine and the phenol being affected considerably.[1]

It has been stated that alkyl phenols brominate quantitatively when the substituting groups are in the *meta*-position or when there are secondary or tertiary groupings in the *ortho-* or *para*-positions. Phenols with primary alkyl groups in the *ortho-* or *para*-positions may give results which are 10–150 % too high.[2] It seems probable, however, that this over-bromination does not take place with cresols or xylenols if the time of reaction is short and if only a slight excess of bromine is used.[3] For accurate results bromination must yield the corresponding bromophenol brom quantitatively, and the subsequent reduction must give the bromophenol completely. The tribromocresol broms are formed only partially (*meta-* is practically theoretical > *ortho-* > *para-*) and reduced incompletely with hydriodic acid (*meta-* theoretical > *ortho-* > *para-*).[4] The values obtained depend very largely on the operator's skill, and of the three cresols only the *meta*-isomer can be estimated accurately by indirect titration although both the *ortho-* and *meta*-compounds may be determined by the direct method. In low concentrations and by using a slightly modified procedure the three cresols can be estimated fairly accurately.[5]

A number of the bromophenol broms are very insoluble in acid, and the precipitate formed in the reaction may contain the fully and partially brominated phenol either in a mixture or as one solid occluding the other. The solid can be dissolved in a solvent, but alcohol must not be used owing to its reaction with bromine.[6] Good results are possible using chloroform which is attacked only slightly.[7] It is essential, however, to cool the reaction mixture before acidifying owing to the high vapour

[1] A. W. Francis and A. J. Hill, *loc. cit.* (p. 87, n. 2).
[2] M. M. Sprung, *loc. cit.* (p. 87, n. 2).
[3] A. W. Francis and A. J. Hill, *Ind. Eng. Chem.* (Anal. ed.), 1941, **13**, 357.
[4] C. M. Pence, *loc. cit.* (p. 87, n. 2); L. V. Redman, A. J. Weith and F. P. Brock, *loc. cit.* (p. 87, n. 2); cf. H. Ditz and F. Cedivoda, *Z. anal. Chem.* 1899, **38**, 873–7; K. K. Järvinen, *ibid.* 1927, **71**, 108–17; W. L. Autenreith and F. Beuttel, *Arch. Pharm.* 1910, **248**, 112–27; S. J. Lloyd, *J. Amer. Chem. Soc.* 1905, **27**, 16–24.
[5] R. D. Scott, *loc. cit.* (p. 87, n. 2).
[6] A. W. Francis and A. J. Hill, *loc. cit.* (p. 87, n. 2).
[7] A. R. Day and W. T. Taggart, *loc. cit.* (p. 87, n. 2).

pressure of the solvent. In the final titration vigorous shaking may be needed near the end-point to prevent traces of iodine being retained in the chloroform layer.

1. *Direct titration with a standard solution of potassium bromate*

A solution of the phenol (25 ml., containing 1·5–2·0 g. per l.) in water or dilute aqueous sodium hydroxide (1 g. for 2·0 g. of phenol) is placed in a flask (500 ml.) and diluted (to 200 ml.). An aqueous solution of potassium bromide (10 ml., 20 %) is added, the mixture is acidified with concentrated hydrochloric acid (5–10 ml.) and titrated with a standard solution of potassium bromate (0·2 N) using starch iodide paper as an external indicator. The end-point is reached when a slight excess of bromine persists for 2–4 min.

2. *Indirect titration*

A solution of the phenol (25 ml., containing 1·5–2·0 g. per l.) in water or dilute sodium hydroxide (1 g. for 2·0 g. of phenol) is placed in a flask fitted with a ground-glass stopper. A slight excess of a standard solution of potassium bromide-potassium bromate is added (25 ml., 0·2 N), the mixture is diluted with water (so that the total volume of water and bromide-bromate solution is 50 ml.) and acidified with concentrated hydrochloric acid (5 ml.). The stopper is replaced quickly, the mixture shaken for 1 min. and then left to stand in a water-bath at $25 \pm 1°$ C. until bromination is completed (about 5 min., see below). The flask is cooled in an ice-bath, potassium iodide (5 ml., 40 %) is added, the mixture is shaken and left for a further 5 min. at $25 \pm 1°$ C. The iodine formed is titrated with a standard solution of sodium thiosulphate (0·1 N), adding starch (1–2 ml., 1 %) near the end-point.

A blank determination with the solvent should be done in parallel.

3. *Indirect titration at low concentrations*[1]

An aqueous solution of potassium bromide (25 ml., 25 %) is added to the phenol (0·02 g.), dissolved in water (200 ml.) or dilute sodium hydroxide (200 ml., 2 %) and placed in a flask

[1] R. D. Scott, *loc. cit.* (p. 87, n. 2).

fitted with a ground-glass stopper. The mixture is acidified with dilute hydrochloric acid (10 ml., 1 : 1) and heated in a water-bath to $25 \pm 1°$ C. A standard solution of potassium bromate (25 ml., containing about 3 g. per l.) is added, and after shaking the flask is placed in a thermostat at $25 \pm 1°$ C. for exactly 1 hr. A solution of potassium iodide (25 ml., 10%) is added, the mixture is shaken and left to stand for exactly 30 min. The iodine formed is titrated with a standard solution of sodium thiosulphate (0·1 N), adding starch (1–2 ml., 1%) near the end-point.

A blank determination should be done as in method 2 above.

PREPARATION: *Potassium bromide-potassium bromate solution* (0·2 N). Potassium bromide (75 g.) and potassium bromate (5·6 g.) are dissolved in water, mixed and the solution is then made up (to 1 l.).

The solution is standardized as described under olefines (p. 15).

Bromination times of various phenols

Phenol	Time of bromination in min.
Phenol	5–30
p-Chlorophenol	30
o-Nitrophenol	30
m-Nitrophenol	5–30
2:4-Dinitrophenol	30
Salicylic acid	30
m-Cresol	1
Resorcinol	1
β-Naphthol	15–20

COLORIMETRIC

A. Gibbs' Method

$$O{=}\langle\!\!\!-\rangle{=}N.Cl + \langle\!\!\!-\rangle OH = O{=}\langle\!\!\!-\rangle{=}N{-}\langle\!\!\!-\rangle{-}OH + HCl.$$

The intensely blue colour of the indophenol solutions was recorded as early as 1835.[1] A number of these compounds were prepared by coupling the *para-*, or more rarely the *ortho-*, reactive hydrogen of phenol with a quinonechloroimide.[2]

[1] H. Robiquet, *Ann. Chem.* 1835, **15**, 289–300; J. Dumas, *ibid.* 1838, **27**, 140–7; R. Kane, *ibid.* 1841, **39**, 39–40.

[2] A. Hirsch, *Ber.* 1880, **12**, 1903–15.

A method of analysis is based on this reaction, and because of its great sensitivity it is used widely for the detection and estimation of small amounts of phenols.[1] A number of chloro- and bromo-quinonechloroimides have been used for coupling and of these either 2:6-dichloro- or 2:6-dibromo-quinonechloroimide is the most suitable. Many substituted phenols give colours with these reagents but relatively few give the characteristic blue colour which is produced with phenol, and then the sensitivity is usually lower. No colour is produced with p-cresol, although o-cresol gives a blue solution. A grouping in the *para*-position usually reduces the sensitivity considerably whilst substitution of the *ortho*-position with a hydrocarbon group has little effect and with halogens in this position the solutions become greenish blue.

Phenol	Colour	Concentration range in parts per billion
Phenol	Blue	5–100
p-Chloro-	Blue	20–400
p-Bromo-	Blue	25–500
o-Chloro-	Green-blue	10–150
o-Bromo-	Green-blue	10–200
2:4-Dichloro-	Green-blue	40–800
2:5-Dichloro-	Green-blue	30–600
2:6-Dichloro-	Green	30–600
Trichloro-	Insensitive	—
Tribromo-	Insensitive	—
o-Phenyl-	Blue	10–200
2-Chloro-6-phenyl-	Blue	10–200
4-Chloro-6-phenyl-	Green-blue	10–200
p-Phenyl-	Insensitive	—
p-tert.-Butyl-	Insensitive	—

2:6-Dichloro- and 2:6-dibromo-quinonechloroimide decompose rapidly in sunlight, and many difficulties which have been reported previously are due to the instability of the reagent. Indophenols act as oxidation-reduction indicators as well as acid-base indicators, so that it is essential to maintain within definite limits both the oxidation potential and the pH.[2] The blue colour develops in alkaline solution, and a pH range

[1] H. D. Gibbs, *J. Biol. Chem.* 1927, **72**, 649–64; A. W. Beshgetoor, L. M. Greene and V. A. Stenger, *Ind. Eng. Chem.* (Anal. ed.), 1944, **16**, 694–6; cf. M. B. Eltinger and C. C. Ruckhoft, *Anal. Chem.* 1948, **20**, 1191–6.

[2] B. Cohen, H. D. Gibbs and W. M. Clark, *Pub. Health Repts.* 1924, **39**, 381 and 804; T. Folpmers, *Chem. Weekblad*, 1934, **31**, 330–3.

between 9·4 and 9·8 should be used.[1] A sodium borate buffer solution can be used, and both the sample and the standard solutions should be made up under identical conditions at the same time. When the pH of the solution is less than 9·4 the colour develops too slowly, and when the value rises above 9·8 the colour may not develop at all. The solutions should be kept in the dark whilst the colour develops, as they become pink if they are exposed to sunlight after the addition of the reagent. As the development of the maximum colour takes a long time all the solutions for comparison should be made up at the same time and kept under the same conditions. The colour is matched either in a colorimeter or in Nessler tubes using a Fisher-Nessler tube support or a similar rack. When a standard curve is prepared for use with a photoelectric absorptiometer care must be taken to maintain a definite period for the production of the colour at a specific temperature, and the same conditions must be used for the preparation of the sample for estimation.[2]

When sulphides are present the colour of the solution seems to depend on the amount of sulphide present and varies from yellow through green to pink. The sulphides must be removed before making up the solution, and lead oxide, lead carbonate and freshly precipitated cadmium carbonate have been used.[3] A more reliable method, however, is to add copper sulphate to both the sample and the standard and filter the solutions.[4] As the estimation is carried out in an alkaline solution calcium and magnesium salts can only be present in very small amounts, otherwise the hydroxides will be precipitated. Interfering salts, turbidity and coloured products can be removed usually by distillation of the phenol from a phosphoric acid solution. When volatile acids, such as salicylic acid, are present two distillations may be necessary. When bases are present sulphuric acid is frequently used instead of phosphoric acid. A modification

---

[1] J. R. Bayliss, *J. Am. Waterworks Assoc.* 1929, **19**, 597–604; E. J. Theriault, *J. Ind. Eng. Chem.* 1929, **21**, 343–6; S. Palitzsch, *Biochem. Z.* 1915, **70**, 333.

[2] I. W. Tucker, *J. Assoc. Official Agr. Chem.* 1942, **25**, 779–82.

[3] R. D. Williams, *J. Ind. Eng. Chem.* 1927, **19**, 530–1; L. S. Renzoni, *J. Am. Waterworks Assoc.* 1940, **32**, 1038.

[4] A. W. Beshgetoor, L. M. Greene and V. A. Stenger, *loc. cit.* (p. 91, n. 1).

which avoids distillation and claims an increased sensitivity
(up to five times) has been described.[1]

An aqueous solution of copper sulphate (1 ml., 0·05 g. of
$CuSO_4$, $5H_2O$ per l.) and a buffer solution of sodium borate
(5 ml., see below) are added to the phenol dissolved in water
(containing between 5 and 100 parts per billion) and to each of
the standards. 2:6-Dibromoquinonechloroimide (1·5 ml. of the
alcoholic solution or 2·0 ml. of the aqueous solution) is added,
the solutions are mixed and left to stand for at least 4 hr. and
preferably overnight. The colour is matched either with a photoelectric absorptiometer or in Nessler tubes using a Fisher-Nessler
tube support or a similar rack.

PREPARATIONS: *Phenol solution to be estimated.* The concentration of the phenol solution to be estimated should be
between 5 and 100 parts per billion, and it may be necessary to
concentrate or dilute the sample.

(a) When the solution contains less than 5 parts per billion.
A measured volume of the phenol solution (containing 5–100 μg.)
is made alkaline with an aqueous solution of sodium hydroxide
(5 ml., 12N) and concentrated (to about 800 ml.). The solution is
cooled and phosphoric acid (10 % solution of 85 % acid) is added
until the pH is between 6·0 and 6·5. The mixture is distilled at
a rate of 8–10 ml. per min. from an all-glass apparatus (fitted
with a splash head) until four samples (of 100 ml. each) have
been collected. The end of the condenser should go well into the
flask to reduce contact with the air as much as possible.

(b) When the solution contains more than 100 parts per billion.
The phenol (500 ml.) and phosphoric acid (0·7 ml., 10 % solution
of 85 % acid) are distilled (as in (a) until 450 ml. have been
collected). Water (50 ml.) is added to the flask and the distillation
is continued (until a total of 500 ml. has been collected). The
distillate is diluted (to 100 ml.) to solutions of different concentrations (no dilution, 1:1, 1:5, 1:10) which are compared with
standard solutions.

With solutions containing more than 10,000 parts per billion
other methods of estimation should be used (see pp. 95–103).

[1] F. C. Basavilbaso, *Rev. obras sanit. nacion* (Buenos Aires), 1947, **11**, no. 117,
158–65.

Standard phenol solutions. A stock solution of phenol is prepared by dissolving pure phenol (1 g.) in water (1 l.) and standardizing by the bromide-bromate method (see pp. 86–90). A series of phenol standards are prepared by diluting the stock solution (1 ml.) with water (to 1 l., see below) and taking 0, 0·5, 1·0, 2·0, 3·0, 4·0, 5·0, 6·0, 8·0, and 10 ml. and making up (to 100 ml.) with water. These standards contain 0, 5, 10, 20, 30, 40, 50, 60, 80, and 100 parts per billion respectively.

The water used to make up the solutions should be free from phenol and chlorine. The phenols can be removed by adding activated charcoal to the water (about 10–20 parts per million is usually enough) and filtering at a pump (through three Whatman papers, no. 40). The water can be freed from chlorine by boiling in an open flask until the volume has been reduced by a quarter.

2 : 6-*Dibromo-* or 2 : 6-*dichloro-quinonechloroimide.* (*a*) Alcoholic solution. A stock solution of 2 : 6-dibromo- (0·1 g.) or 2 : 6-dichloro-quinonechloroimide (0·7 g.) in ethyl alcohol (25 ml., 95 %) can be stored in a small dark brown, glass-stoppered bottle for 3 or 4 days if it is kept in a cool place. The reagent is prepared immediately before use by diluting (5 ml.) with distilled water (to 100 ml.), preferably in a dark room or in a dark brown bottle.

(*b*) Aqueous solution. 2 : 6-Dibromo- (0·04 g.) or 2 : 6 dichloro-quinonechloroimide (0·03 g.) is ground in a mortar with water (10 ml.). The fine suspension is washed into a brown glass-stoppered bottle, diluted (to approximately 100 ml.) with water, shaken for 10 min. and filtered. This solution decomposes very rapidly and must not be used after 20 min.

Sodium borate buffer solution. Anhydrous sodium tetraborate (15 g.) is dissolved in warm distilled water (900 ml.), sodium hydroxide is added (3·27 g. in 80 ml. of water), and the solution is made up (to 1 l.). A pH of 9·6 ± 0·3 should result from adding 5 ml. of this solution to 100 ml. of the aqueous phenol solutions.

B. Coupling with Diazo Compounds

$$C_6H_5.OH + Cl.N_2R' = R'\!\!-\!\!N\!\!=\!\!N\!\!-\!\!\langle\overline{}\rangle\!\!-\!\!OH$$

An alkaline solution of a phenol reacts with diazo compounds to form solutions which are strongly coloured even at high dilution. Diazotized sulphanilic acid[1] or p-nitroaniline[2] have been used for the estimation of phenols. With sulphanilic acid the colours vary from greenish yellow through orange to red, and with p-nitroaniline a very stable yellow-orange solution is obtained. With the former the solution colours are greenish yellow with phenol, yellow with o- and m-cresol, and red with compounds in which the $para$-position is blocked and coupling takes place with the $ortho$-hydrogen (p-cresol, p-hydroxyphenylacetic acid, p-hydroxyphenylpropionic acid and p-hydroxyphenyllactic acid). Tyrosine and tyromine give pink solutions which change rapidly to yellow and then fade. The colours can be matched against standard solutions made at the same time under identical conditions or against artificial standards, such as a platinum-cobalt[3] or an Arny-Ring standard.[4]

A micro-estimation method for concentrations as low as one part in four million has been described for phenol, o-, m- and p-cresols, p-hydroxyphenyllactic acid, and p-hydroxyphenyl-propionic acid.[5]

(a) *With sulphanilic acid.* A freshly prepared aqueous solution of sodium nitrite (2 ml., 8 %) is added to a mixture of sulphanilic acid (4 ml., see below) and the phenol (50 ml., containing approximately 1 mg. per l.). The mixture is shaken

[1] J. N. Miller and M. Urbain, *Ind. Eng. Chem.* (Anal. ed.), 1930, **2**, 123–4 M. T. Hanke and K. K. Koessler, *J. Biol. Chem.* 1922, **50**, 235–70; J. J. Fox and A. J. H. Gange, *J. Soc. Chem. Ind.* 1920, **39**, 260T.

[2] R. C. Theis and S. R. Benedict, *J. Biol. Chem.* 1924, **61**, 67–71; H. Bach, *Gas Wasserfach*, 1929, **72**, 375–7; A. G. Nolte, *Chem. Ztg.* 1933, **57**, 654; A. D. Marenzi, *Anales farm. bioquim.* (Buenos Aires), 1931, **2**, 187–90; *Compt. rend. soc. biol.* 1932, **109**, 321–2; *ibid.* 1932, **110**, 147–8; *Rec. soc. argentina biol.* 1932, **8**, 26–37; R. F. Banfi and A. D. Marenzi, *ibid.* 1935, **11**, 509–18, and *Compt. rend. soc. biol.* 1935, **120**, 812–14; cf. W. Schuler and P. Heinrich, *Experientia*, 1945, **1**, 235; H. Friend, *J. Biol. Chem.* 1923, **57**, 497–505.

[3] A. Hazen, *Am. Chem. J.* 1892, **12**, 427–8.

[4] H. V. Arny and C. H. Ring, *J. Franklin Inst.* 1915, **180**, 199–213; H. V. Arny and A. Taub, *ibid.* 1923, **196**, 858.

[5] M. T. Hanke and K. K. Koessler, *loc. cit.* (p. 95, n. 1).

and an aqueous solution of sodium hydroxide (5 ml., 10 %) is added, mixed and left for 3 min., when the full colour develops. A comparison is made with standard solutions or with artificial colours which have been standardized previously.

(b) *With p-nitroaniline.* A diazotized solution of *p*-nitroaniline (1 ml., see below) is added to a mixture of the phenol (10 ml., approximately 0·0025 mg./ml.) and gum acacia (1 ml., 1 %). After 1 min. an aqueous solution of sodium carbonate (2 ml., 20 % sodium carbonate) is added and the liquid diluted to a known volume. Standard solutions are prepared at the same time under identical conditions. A comparison of the colours is made after 2–4 min.

PREPARATIONS: *Sulphanilic acid.* Concentrated sulphuric acid (1 ml.) is added to sulphanilic acid (20 g., recrystallized) dissolved in water and the solution is made up (to 250 ml.).

A diazotized solution of p-nitroaniline. *p*-Nitroaniline (1·5 g.) is added to water (50 ml.) and dissolved by adding concentrated hydrochloric acid (40 ml.). The solution is diluted (to 500 ml.), and immediately before use it is diazotized by adding (to 25 ml.) a freshly prepared aqueous solution of sodium nitrite (0·75 ml., 10 %).

The diazo reagent can be kept for 1 day.

Standard phenol solutions. A stock solution of the phenol (0·1 %) is made up in dilute hydrochloric acid (0·1 N) and is used for the preparation of the standard solutions which should be made up immediately before use. For a concentration of 1 mg. per l. the phenol solution (10 ml., 0·1 %) is diluted (to 1 l.) with water. For a concentration of 0·0025 mg. of the phenol per ml. the phenol solution (10 ml., 0·1 %) is diluted (to 100 ml.) with water and then diluted again (2·5–100 ml.).

C. With Millon's Reagent

Phenols react with nitrous or nitric acids, alone or in the presence of metallic salts, to form highly coloured products. Substitution usually takes place in the *para*-position, but even when this is blocked a colour is produced although it may develop more slowly. Mercury salts accelerate the rate of reaction,[1] and Millon's reagent,[2] which is a solution of mercury in a mixture of nitrous and nitric acids, may be used for the estimation of phenols. The intensity of the colour can be increased either by using a relatively large amount of the reagent or by heating the solution, but estimation is possible only by a comparison of solutions made up at the same time under identical conditions. The colour varies from a deep red with phenol to a greenish yellow with p-cresol, but traces of impurities may change the colour. The addition of a small amount (0·1 mg.) of m-cresol to a solution of phenol changes the colour from red to orange. The preparation of the reagent is important as an excess of nitric acid causes the colour to fade quickly, but if too little acid is used the normal colour of certain phenol solutions persists too long. Not more than a trace of alcohol should be present and the concentration of hydrochloric acid must be kept below 0·33 %, otherwise the colour is not produced.[3]

The procedure using Nessler tubes is complicated and tedious and frequently the results are poor when orange colours have to be matched, as with the cresols. A quicker method of estimation uses a photoelectric absorptiometer for the comparison and so avoids the preparation of the blank solutions with formaldehyde.[4] The method has been used for the estimation of phenols in sewage,[5] urine,[6] serum,[7] glycerine,[8] tar acids,[9] and phenol in the presence of other phenols.[10]

1. Using Nessler tubes

Two pairs of tubes (graduated for 25 ml.) are made up with equal volumes of the phenol solution to be estimated (containing

[1] H. D. Gibbs, *J. Biol. Chem.* 1927, **71**, 445–59.

[2] M. Th. Kohs, *Pharm. Weekblad*, 1931, **68**, 557–69.

[3] G. Haas and W. Trautmann, *Z. physiol. Chem.* 1923, **127**, 52–66.

[4] T. S. Harrison, *J. Soc. Chem. Ind.* 1943, **62**, 119–23; *ibid.* 1944, **63**, 312–3.

[5] H. Bach, *Z. anal. Chem.* 1911, **50**, 736–40.

[6] M. Weisz, *Biochem. Z.* 1920, **110**, 258–65.

[7] E. Elvove, *U.S. Pub. Health Service, Hyg. Lab. Bull.* 1917, **110**, 25–33.

[8] G. Denigès, *Bull. soc. pharm. (Bordeaux)*, 1926, **64**, 3–15; *ibid.* 1927, **65**, 118–20.

[9] T. S. Harrison, *loc. cit.* (p. 97, n. 4).

[10] R. M. Chapin, *J. Ind. Eng. Chem.* 1920, **12**, 771–5; *U.S. Dept. of Agr. Bull.* 1924, p. 1308.

about 4 mg.) in one pair and equal volumes of a standard solution (see below) in the other pair. The reagent (5 ml.) is added down the side of each tube, the liquids are mixed and heated in a boiling water bath for 30 min. and then cooled in cold water for 10 min. The solutions are acidified with nitric acid (5 ml., 1:4 free from oxides of nitrogen) and mixed thoroughly. Formaldehyde (3 ml., 37 %, commercial, diluted to 100 ml. with water) is added to one tube from each pair, all four liquids are made up (to 25 ml.), a stopper is placed in each tube, the contents are mixed and left overnight. The solutions containing formaldehyde fade to a yellow colour. If a precipitate forms it should be filtered off 10 min. after the solid separates and the filtrate used in the estimation. A measured volume (20 ml.) from each of the two standard solutions is placed in a graduated flask (100 ml.), acidified with nitric acid (5 ml., 1:4, free from oxides of nitrogen), mixed and used to fill two burettes. Equal volumes (10 ml.) from the other two solutions are transferred to two Nessler tubes. One solution should be yellow and the other one deeply coloured. Measured volumes are run in from the burettes: the deeply coloured standard into the yellow and the standard bleached with formaldehyde into the coloured solution to be estimated. The volumes added are adjusted until the depth of colour in the two tubes is matched.

A simpler method has been used for cresols.[1] The cresol (10 ml.) and the reagent (3 ml.) are mixed and heated at 100° C. for 5 min. The liquid is filtered and compared with a standard solution of a cresol prepared in parallel. This method is less accurate than that above.

2. *Using a Spekker photoelectric absorptiometer.*

The solution is made up as above, and a reading is taken in a Spekker photoelectric absorptiometer. An Ilford yellow-green filter (no. 605, 4 cm. cell) is used for most phenols (a violet filter, no. 601, for m-cresol). The amount of phenol can be determined either by comparison with standard solutions or by using a graph prepared previously from standard solutions.

[1] E. V. Alekseevskii and K. G. Tarasova, *J. Appl. Chem.* (U.S.S.R.), 1935, **8**, 1313–18.

PREPARATIONS: *Millon's reagent.* Concentrated nitric acid (20 ml.) is added to mercury (2 ml.) in a flask (100 ml.) and left to stand in a fume cupboard until the mercury has dissolved (about 15 min.). Water (35 ml.) is added, and if the basic salt separates from the solution concentrated nitric acid is added drop by drop until a clear solution is obtained. An aqueous solution of sodium hydroxide (10 %) is added with mixing, a drop at a time until a permanent cloudiness remains. Dilute nitric acid (5 ml., 1:4) is added and the mixture shaken.

A fresh supply of the reagent should be made up each day.

The nitric acid used in the preparation of the reagent and in making up the solutions should be freed from oxides of nitrogen by drawing through it a slow stream of dry air. If the acid in the reagent decomposes slightly on standing the oxides of nitrogen must be removed immediately before use.

Standard solution of a phenol. A stock solution of phenol, or a cresol (1 g.), is made up in water and diluted (5 ml. to 200 ml.) immediately before it is used as a standard.

D. Formation of Nitrosophenols with Nitrous Acid

Nitrosophenols in alkaline solution give highly coloured solutions which can be used for the estimation of relatively high concentrations of phenols (about 0·2 mg. compared with 5 parts per billion by the Gibbs method.[1] A mixture of nitric and sulphuric acids at about 100° C. was used originally, but a simpler procedure for the preparation of the nitrosophenols at room temperature is possible using a solution of sodium nitrite instead of nitric acid. The colours of the alkaline solutions range from greenish yellow to orange-yellow, and for accurate results comparison should be made with a standard solution of the same phenol. When the order of concentration only is needed an arbitrary standard of phenol or a cresol is chosen. The method is unsatisfactory for compounds giving coloured solutions and

---

[1] R. W. Stoughton, *J. Biol. Chem.* 1936, **115**, 293–8; L. Lykhen, R. S. Treseder and V. Zahn, *Ind. Eng. Chem.* (Anal. ed.), 1946, **18**, 103–9; L. A. Wetlaufer, F. J. Van Natta and H. B. Quattlebaum, *ibid.* 1939, **11**, 438–9; J. Rae, *Pharm. J.* 1927, **119**, 332; 1930, **124**, 239; H. D. Gibbs, *J. Biol. Chem.* 1927, **71**, 445–59.

for hydroxybenzoic acids, hydroxyphenyl aldehydes and ketones and their derivatives owing to the development of a very faint colour.

Two alternative procedures are described, as it is sometimes necessary to extract the phenol before estimation. Aniline and xylidene interfere and must be removed completely by shaking with acid. The method can be used for the estimation of mono- and di-hydroxyphenols, naphthols and alkyl phenols in petroleum, kerosene, fuel oil, hydrocarbon solvents, naphthenic acids, disinfectants, pharmaceutical preparations, aqueous solutions and similar materials. A method for the micro-determination of phenol in blood plasma has been described.[1]

1. *Direct method*

This method can be used for phenols, and materials containing phenols, which are soluble in an acetic acid buffer solution. These include cresylic acids, alcohols, naphthenic acids, glycerols, phenyl ethers, ketones and aqueous solutions of phenols. Highly coloured and opaque solutions must be extracted before the estimation is done.

The phenol (3–8 mg.) is dissolved in an acetic acid buffer solution (see below) and made up (to 100 ml.) in a graduated flask. The solution is mixed and a measured volume is transferred to a flask (50 ml.). Sulphuric acid (5 drops, 36 N) and an aqueous solution of sodium nitrite (2 ml., saturated) are added gradually with continuous shaking. The liquid is cooled in an ice-bath for 15–30 min. (or 30–45 min. for phenol itself), and alcoholic ammonia (see below) is added slowly (to bring the total volume to 500 ml. exactly). The mixture is left for 1 hr., or, if possible, overnight, and estimated either by reading in a photoelectric colorimeter using violet light (420 mμ) or by comparison with standard solutions.

When a photoelectric colorimeter is used three blank solutions should be made up in parallel with the phenol to be estimated. The sodium nitrite solution is omitted from the first and twice the volume of the phenol solution is added; phenol is left

[1] J. García-Blanco and F. Comesaña, *Anales soc. españ. fis. quím.* 1932, **30**, 690.

out of the second and neither the phenol nor sodium nitrite are added to the third.

2. *Estimation after extraction of the phenol.*

When the sample is insoluble in the acetic acid buffer solution or when the solution is highly coloured or opaque the phenol must be extracted first with an aqueous solution of potassium hydroxide.

The sample containing the phenol (up to 10 mg. of the phenol) is placed in a separating funnel (about 125 ml.) and diluted (to 50 ml.) with a suitable solvent (see below). An aqueous solution of potassium hydroxide (10 ml., 10 %) is added, the mixture is shaken for 5 min. and left to stand. Two layers separate and the lower one is run off completely into a flask (100 ml.). The extraction is repeated with aqueous potassium hydroxide (5 ml., 10 %) and finally with water (5 ml.). The alkali and water extracts are combined, cooled in an ice-bath and made up (to 100 ml.) by slowly adding an acetic acid buffer solution (see below). A measured volume (1–5 ml., containing 0·05–0·5 mg. of the phenol) is transferred to a dry flask (50 ml.), an acetic acid buffer solution is added (to bring the total volume to 5 ml.) and the estimation is then carried out as described above from the addition of sulphuric acid (5 drops, 36 N).

3. *Estimation after extraction of aniline and xylidene*

Aniline and xylidene interfere to some extent and the phenol cannot be estimated by the direct method, but satisfactory results are possible after extraction, provided the concentration of the phenol is less than twenty times greater than the aniline or xylidene concentration. For samples having a higher relative concentration the aromatic bases are extracted with sulphuric acid.

A sample of the phenol (containing approximately 15 mg. of the phenol) is weighed into a separating funnel and dissolved in technical octane (100 ml., phenol-free). Sulphuric acid (10 ml., 5 %) is added and the solution is shaken vigorously for 5 min. Two layers form on standing, the lower one is run off into a flask and the top one is shaken with water (10 ml.) for

1 min. The acid and aqueous extracts are combined, and any hydrocarbon which separates is removed and added to the octane solution. The estimation is then carried out as described under (2).

Accuracy ± 1 % (with a photoelectric absorptiometer); ± 2 % by other methods.

PREPARATIONS: *Standard solution of a phenol.* Phenol (0·1–0·01 g.) is weighed into a graduated flask (100 ml.) and made up (to 100 ml.) with the acetic acid buffer solution. The liquid is mixed and a measured volume (10 ml.) is diluted (to 100 ml.) with the buffer solution.

The phenol used in the preparation of the standard should if possible be the same as the one to be estimated.

Acetic acid buffer solution. Potassium hydroxide (150 ml., 10 %) is added to glacial acetic acid (800 ml.) and the solution is made up (to 1 l.) with water (50 ml.).

Alcoholic ammonia solution. Dry ethyl, or isopropyl alcohol (450 ml.) is added to an aqueous solution of ammonia (300 ml., 14N) and water (250 ml.).

The solvent. A suitable solvent, such as technical octane, benzene, toluene or ether (1 l.), is extracted with an aqueous solution of potassium hydroxide (100 ml., 10 %), washed with water (100 ml.) and filtered through a dry filter paper.

E. With the Folin-Denis Reagent

Phenols can be estimated with a mixture of phosphotungstic and phosphomolybdic acids, which is known usually as the Folin-Denis reagent (cf. pp. 124–5).[1] When the mixture is reduced an intense blue solution is formed which contains a tungstic oxide. The method cannot be used in the presence of a reducing agent, and as it is often almost impossible to remove the last trace of these interfering substances the values obtained always tend to be a little high. Ammonium sulphide, hydrogen sulphide, sulphurous acid, hydriodic acid, stannous and ferrous salts, hydrogen peroxide, unsaturated aliphatic compounds, amino compounds, urea, uric acid, haemoglobin and reducing

[1] O. Folin and W. Denis, *J. Biol. Chem.* 1915, **22**, 305–8.

sugars all interfere and must be removed as completely as possible before the development of the colour.[1] Proteins can be removed either with colloidal iron or a cream of alumina,[2] and the solution can then be freed from uric acid with silver lactate.[3]

A rapid method for micro-determination of phenol in blood has been described by which a large number of determinations can be made in a short time.[4] The low concentration of phenol (about 2·5 mg. in 100 ml.) limits the accuracy of the method to + 5 %.

The reagent (5 ml.) and an aqueous solution of sodium carbonate (15 ml., 20 %) are added to a solution of the phenol in water (containing 0·5 mg. of the phenol). The mixture is made up to a definite volume with water warmed to 30–35° C., and after standing for 20 min. it is estimated either by comparison with standard solutions prepared at the same time under identical conditions or by reading in a Spekker photoelectric absorptiometer (Ilford red filter no. 608, 4 cm. cell, tungsten filament lamp).[5]

PREPARATIONS: *Folin-Denis reagent.* Phosphoric acid (50 ml., 85 %) and concentrated hydrochloric acid (100 ml.) are added to water (700 ml.), sodium tungstate (100 g.) and sodium molybdate (25 g.) in a Quickfit flask (1500 ml.). The mixture is boiled gently under a reflux condenser for 10 hr. Lithium sulphate (150 g.), water (50 ml.) and bromine (a few drops) are added and the flask is heated without a condenser for 15 min. to remove the excess bromine. The liquid is cooled, diluted (to 1 l.) and filtered. If the filtrate is a greenish colour reducing products are present and must be removed by adding more bromine (1–2 drops) and reheating without a condenser. The reagent should be kept in a bottle and protected against dust.

Phenol solution. See p. 94.

[1] E. Schreiner, *Biochem. Z.* 1929, **205**, 245–55; H. Fujiwara and E. Katoaka, *Z. physiol. Chem.* 1933, **216**, 133–7.
[2] K. F. Pelkan, *J. Biol. Chem.* 1922, **50**, 491–7.
[3] Y. Asada, *Tôhoku J. Exp. Med.* 1930, **15**, 363–8.
[4] *Ibid.*
[5] T. S. Harrison, *J. Soc. Chem. Ind.* 1944, **63**, 312–13.

Miscellaneous

(1) Iodine can be used in the same way as bromine (see pp. 86–90) for the estimation of phenols.[1] An excess of a standard solution of iodine is added to an alkaline solution of the phenol, and after the reaction is finished the excess is determined by titration with sodium thiosulphate. The results are usually less accurate than those found by the bromide-bromate method (see pp. 89–90). Many modifications of the original procedure have been suggested, including dissolving the phenol in sodium bicarbonate instead of sodium hydroxide[2] and the use of a buffer solution to keep the hydrogen-ion concentration low.[3]

(2) Phenols react with chloramine-T to give coloured solutions which can be used for their estimation. Comparison should be made with a standard solution of the same phenol. Phenol itself gives no colour.[4]

(3) Low concentrations of a phenol may be determined by adding bromine water and comparing the turbidity with that produced in a standard solution.[5] The phenol is separated from contaminating material by distillation or extraction. Alcohols, amines, amino compounds, aldehydes and oils react with bromine water and so interfere. An accuracy of about 10 % is possible with concentrations as low as 1 part in a million, and with stronger solutions better results are possible.

(4) An indophenol is formed when a phenol is oxidized with sodium hypochlorite in the presence of p-dimethylphenylene-diamine. The highly coloured reaction product is extracted with carbon tetrachloride and compared with a standard prepared under the same conditions. The extraction increases the sensi-

[1] J. Messinger and G. Vortmann, Ber. 1890, 23, 2753–6; F. W. Skirrow, J. Soc. Chem. Ind. 1908, 27, (2), 58–63; H. J. Rose and F. W. Sperr, Am. Gas Assocn. Monthly, 1920, 2, 117–20; L. V. Redman, A. J. Weith and F. P. Brock, J. Ind. Eng. Chem. 1913, 5, 831–6.

[2] Ibid.

[3] J. d'Ans, Z. anal. Chem. 1934, 96, 1–6.

[4] B. N. Afanas'ev, Khim. Prom. 1944, no. 7, 18–19; J. Appl. Chem. (U.S.S.R.), 1944, 17, 335–6.

[5] J. A. Shaw, Ind. Eng. Chem. 1929, 21, 118–21; ibid. 1931, 3, 273–4; F. W. Skirrow, loc. cit.; F. Meinch and M. Horn, Angew. Chem. 1934, 47, 625–8.

tivity of the method and it is essential when the phenol concentration falls below 15 parts in ten million.[1]

(5) The precipitation of the insoluble bromo- and iodophenols has been used as a gravimetric method of estimation. The iododerivatives do not always have the expected composition, but there is a definite relationship between the weight of the precipitate and the original weight of the phenol.[2]

(6) The method of estimation described under alcohols can also be used for phenols (see p. 69, paras. (4) and (5)).

(7) A modification of Koppeschaar's method (see pp. 86–90) has been described in which the excess bromine is titrated with sodium arsenite (0·1 N) or sodium thiosulphate (0·1 N). It is claimed that the method gives accurate results for *ortho*- and *para*-cresols.[3]

(8) Phenols may be determined colorimetrically by their reaction with copper sulphate in a solution containing ammonia and hydrogen peroxide.[4]

(9) Phenols react with titanium tetrachloride in the absence of water to form a red solution of $C_6H_5O.TiCl_3$, which can be used for their colorimetric estimation. Water, alcohols and ethers interfere.[5]

(10) Isomeric cresols and cresol-phenol mixtures can be detected and estimated by comparing their ultra-violet absorption spectra with standard spectrograms using pure phenols.[6]

(11) A reaction mixture of iodine and silver nitrate in alcohol has been used for the estimation of phenols. An excess of the reagent is added to the phenol, and when the reaction is finished

[1] G. U. Houghton and R. G. Pelly, *Analyst*, 1937, **62**, 117–20; cf. L. F. Levy, *J. S. African Chem. Inst.* 1939, **22**, 2936.

[2] *Sampling and Analysis of Coal, Coke and By-products*, Carnegie Steel Company, 1929, p. 153; M. François and L. Seguin, *Bull. soc. chim.* 1933, **53**, 711–23.

[3] W. Bielenberg, H. Goldhahn and A. Zoff, *Oel u. Kohle*, 1941, **37**, 496–500; H. Bach, *Gas u. Wasserfach*, 1931, **74**, 331–4.

[4] G. Denigès, *Bull. soc. pharm. (Bordeaux)*, 1931, **69**, 233–6.

[5] G. P. Luchinskii, *Zavodaskaya Lab.* 1936, **5**, 233–4.

[6] W. W. Robertson, N. Ginsberg and F. A. Matsen, *Ind. Eng. Chem.* (Anal. ed.), 1946, **18**, 746–50.

the unchanged iodine is titrated with a standard solution of sodium thiosulphate.[1]

(12) A solution of a phenol (approximately 0.1N), acidified with hydrochloric acid (4N), may be titrated with iodine mono-chloride using starch iodide as an indicator. The method is successful with phenol, resorcinol, β-naphthol and 8-hydroxy-quinoline but with salicylic acid and zinc sulphophenolate the reaction is too slow for accurate results.[2]

(13) A mixture of formaldehyde and the phenol is heated and the polymer which is formed is filtered, washed, dried at 140° C. and weighed.[3]

[1] Ya. A. Fialkov and A. I. Gengrinovich, *Zapiski Inst. Khim. Akad. Nauk.* (U.S.S.R.), 1940, 7, 125–38; *J. Gen. Chem.* (U.S.S.R.), 1941, 11, 596–604.
[2] A. I. Gengrinovich, *Farniatsiya*, 1947, 10, no. 2, 23–8.
[3] M. Krahl, *Kunstoff- Tech. u. Kunstoff-Anwend*, 1942, 12, 190–1.

CHAPTER III

MERCAPTANS

The mercaptans are neutral compounds, but although they are the sulphur analogues of alcohols they form metallic derivatives easily and rapidly. Many of the methods which have been described for their estimation use this property as is shown by the following reactions:

Estimation by silver nitrate:

$$R.SH + AgNO_3 = R.S.Ag + HNO_3.$$

    A. Addition and back titration of an excess of silver nitrate.

    B. Electrometric titrations.

Estimation by the cupric salt of an organic acid:

$$4R.SH + 2Cu(OAc)_2 = 2CuSR + R.S.S.R + 4HOAc.$$

Indirect acid titration:

$$2R.SH + Pb(OH)^+.CH_3.COO^- = \begin{matrix} R.S \\ R.S \end{matrix}\!\!>\!Pb + H_2O + CH_3.COOH,$$

$$\begin{matrix} R.S \\ R.S \end{matrix}\!\!>\!Pb + H_2SO_4 = 2R.SH + PbSO_4.$$

Oxidation with iodine:

$$2R.SH + I_2 = R.S.S.R + 2HI.$$

Colorimetric.

Miscellaneous.

ESTIMATION BY SILVER NITRATE

$$R.SH + AgNO_3 = R.S.Ag + HNO_3.$$

Mercaptans react easily and quantitatively with silver nitrate forming the corresponding silver salt and they can be estimated by the following procedures.

A. Addition and Back Titration of an Excess of Silver Nitrate

A solution of the mercaptan, either in petrol or in some other hydrocarbon solvent, is shaken with an excess of silver nitrate, which latter is determined by titration with ammonium thiocyanate using ferric alum as the indicator.[1] An emulsion is formed usually but it can be broken down by the addition of a few ml. of methyl or ethyl alcohol. High values may result from occlusion of silver ions in the small lumps of the mercaptide formed on the addition of silver nitrate, and the solid may pass into the hydrocarbon layer carrying with it absorbed silver nitrate. It has been suggested that the excess of silver nitrate should be estimated by adding ammonium thiocyanate in excess, shaking the mixture in a machine and then doing a final titration to find the ammonium thiocyanate. This often results in low values if the thiocyanate ion is held in the emulsion or scum which is formed. Accurate values are possible only when several determinations are made and the average taken of at least three readings which differ only by an experimental error. Unless great care is taken high values are obtained owing to the great difficulty of removing the excess of silver nitrate held in the emulsion. The estimation of coloured mercaptans or oils sometimes proves difficult, as the end-point is masked by the colour of the original solution. It is usually necessary to separate the oil, filter off the silver mercaptide, wash the precipitate and titrate the solution and the washings.

The method has been used widely for the easy estimation of aliphatic mercaptans in petrol, but thiophenols give low results owing to the formation of the corresponding sulphides as by-products. Erratic values have been recorded with this method, but it is reported that consistent results are possible when the concentration of the standard silver nitrate solution is reduced from 0·05N, as used originally, to 0·005N,[2] and the

---

[1] P. Borgstrom and E. E. Reid, *Ind. Eng. Chem.* (Anal. ed.), 1929, **1**, 186–7; W. M. Malisoff and E. M. Marks, *J. Ind. Eng. Chem.* 1931, **23**, 1114–20; W. M. Malisoff and C. E. Anding, Jr., *Ind. Eng. Chem.* (Anal. ed.), 1935, **7**, 86–8; J. S. Ball, *U.S. Bur. Mines, Rept. Investigations*, 1942, p. 3591; G. E. Mapstone, *Australian Chem. Inst. J. and Proc.* 1946, **13**, 373–7.

[2] W. M. Malisoff and C. E. Anding, Jr., *loc. cit.* (p. 108, n. 1).

concentration of sulphur present as mercaptan does not exceed 0·075–0·1 %.

Free sulphur is present frequently in petroleum fractions and may be removed by shaking with mercury. The estimation is unaffected by mercury itself, but absorption can take place on the metallic sulphide and so give low results.[1] Consistent and accurate values are possible without removing the sulphur, although the titration may be more difficult to follow as the solution may pass through varying shades of orange, red or brown depending on the amount of sulphur present. It is usually very difficult to break down the emulsion under these conditions, and it may be necessary to add a large volume (up to 100 ml.) of alcohol to do this and bring the precipitate into the interface so that the colour can be seen.

Methyl alcohol (5–10 ml., 95 %) and silver nitrate (15–30 ml., excess, 0·005 N) are added to a petrol solution of the mercaptan (about 0·075 % sulphur) in a glass-stoppered bottle (16 oz.). The mixture is shaken vigorously, ferric alum (2 ml., see below) is added and the excess silver nitrate determined by titration with ammonium thiocyanate (0·005 N) until a faint pink colour persists. A small excess of silver nitrate (5 ml., 0·005 N) is added, the titration with thiocyanate repeated and this reading taken as the end-point. Constant shaking is necessary during the titrations and care should be taken not to confuse the end-point with the slight tint due to the indicator.

Percentage of mercaptan sulphur

$$= \frac{\text{ml. of AgNO}_3 \times \text{normality of AgNO}_3 \times 0\cdot03206 \times 100}{\text{weight of mercaptan}}.$$

Accuracy ± 1·5 % (4 %).

PREPARATION: *Ferric alum*. Ferric alum (40 g.) is dissolved in water (70 ml.), nitric acid (20 ml., 6 N) is added and the solution made up (to 100 ml.). The indicator is prepared immediately before use by boiling the solution (1 part) to remove any oxides of nitrogen and diluting with water (3 parts).

[1] R. T. Bell and M. S. Agruss, *Ind. Eng. Chem.* (Anal. ed.), 1941, **13**, 297–9; H. Schindler, G. W. Ayres and L. M. Henderson, *ibid.* 1941, **13**, 326–8; cf. E. J. Greer, *J. Ind. Eng. Chem.* 1929, **21**, 2033.

B. Electrometric Titrations

1. *Potentiometric titration in alcoholic solution*

The titration of a mercaptan with silver nitrate in homogeneous solution stops the precipitated mercaptide passing into a second phase carrying absorbed silver nitrate (see above). This is done by dissolving the mercaptan in either alcohol[1,2] or alkali[3] and titrating with silver nitrate. The end-point is determined electrometrically using a fixed[1,3] or rotating[2] indicator electrode.

When the titration is done in alcoholic solution with a fixed electrode sufficient alcohol is placed in the cell to dissolve the hydrocarbon or dilute the solution of the mercaptan. An alcoholic solution of silver nitrate is used, and as the end-point is determined by a change in potential of a polished silver electrode, it is independent of the colour of the original solution (see above). The silver mercaptides are sparingly soluble and therefore substances which normally react with silver nitrate but form compounds more soluble than the silver mercaptides do not interfere. Silver sulphide, however, is less soluble than the mercaptides, so that hydrogen sulphide and sulphur should be removed first from the mercaptan solution by shaking successively with acidified cadmium sulphate and mercury.

The potentiometric titration in aqueous-alkaline solution is similar to the above. The indicator electrode is a silver rod either polished or coated electrically with a thin layer of silver sulphide, which latter gives a greater potential drop at the end-point. The mercaptan solution is dissolved in aqueous sodium hydroxide, and ammonium hydroxide is added to prevent the co-precipitation of silver oxide with the insoluble mercaptide. Very few substances interfere owing to the relatively high potential at which the mercaptides are precipitated, but the presence of water-soluble sulphur compounds which form salts insoluble

[1] M. W. Tamele and L. B. Rylands, *Ind. Eng. Chem.* (Anal. ed.), 1936, **8**, 16–19.
[2] I. M. Kolthoff and W. E. Harris, *ibid.* 1946, **18**, 161–2; cf. J. T. Stock, *Metallurgia*, 1947, **36**, 51–4.
[3] M. W. Tamele, L. B. Rylands and V. C. Irvine, *Ind. Eng. Chem.* (Anal. ed.) 1941, **13**, 618–22.

in ammonium hydroxide may give trouble. The presence of sulphide and thiosulphate ions can be recognized readily from the titration curves (see fig. 20), but between certain limits mercaptans can be estimated quantitatively with both of these ions present. When thiosulphate ions are in solution the end-point is hardly detectable using a polished silver electrode, but the change in potential is much greater and more distinct with a sulphide-coated rod.

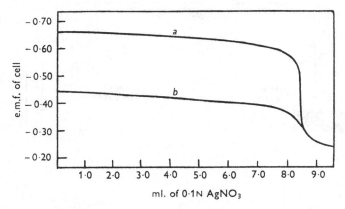

Fig. 20. *a*, Titration with a sulphide-coated electrode; *b*, titration with a polished silver electrode. Sulphide ions are precipitated at a more negative and thiosulphate at a more positive potential than mercaptans.

In alkaline solution mercaptans are oxidized readily to disulphides and other compounds which may cause serious errors, far greater than those inherent in the experimental conditions. The alkaline solution should not be exposed needlessly to the air, and the titration should be done immediately after the mercaptan has been added. Accurate control of the pH of the solution is important, as below pH 13 serious evaporation losses take place owing to the presence of undissociated mercaptan, while above pH 14 co-precipitation of silver oxide and the mercaptide may take place and high values result. The most accurate results are obtained when a normal solution of sodium hydroxide is used. Silver oxide is precipitated if insufficient ammonium hydroxide is added to the titration cell, and although this cannot be detected easily from the shape of the curve it is shown clearly in the colour of the precipitate. The colour of the silver mercaptides changes from white to light yellow as the molecular weight increases, and co-precipitation results in a distinct and easily noticeable darkening of the solid.

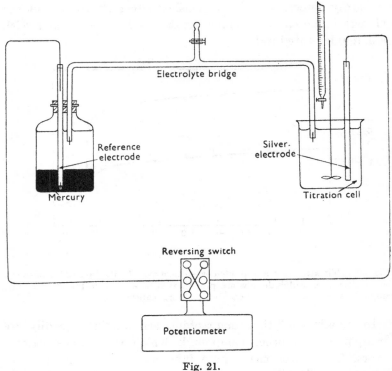

Fig. 21.

The apparatus is as shown in fig. 21. The silver half-cell consists of an electrode of polished silver wire (about 2 mm. in diameter) immersed in an alcoholic solution of sodium acetate (50 ml., see below). The electrolyte bridge and the mercury half-cell are filled with the same solution. The positive electrode is a layer of mercury (3–4 mm. deep), and in the absence of a generally accepted standard reference electrode for alcoholic solutions, the mercury half-cell is taken as the standard and the potential of the silver electrode is considered equal to the numerical value of the e.m.f. in the cell. The values developed are fairly constant (-0.070 volt, the silver wire is the negative electrode) although variations may be caused by impurities in the silver wire (see below). The cell is

Ag⁻	Alcoholic solution of sodium acetate 0·1N	Alcoholic solution of sodium acetate 0·1N	Hg⁺.

The resistance of the cell is high, so that a fairly sensitive potentiometer must be used.

A solution of the mercaptan (free from hydrogen sulphide and elementary sulphur, see below) in kerosene, petrol, amylene, or an aliphatic alcohol is added to the alcoholic sodium acetate solution. The amount of mercaptan present may vary with the solubility in ethyl alcohol (from 5 to 10 ml. of a hydrocarbon solution is soluble in the cell liquid), and the percentage of mercaptan in the solution (a titration value of 10–15 ml. of 0·01N silver nitrate is aimed at normally). The potential of the silver electrode rises from -0.070 to -0.038 volts. A standard solution of silver nitrate (0·01N) in *iso*propyl alcohol is added in small amounts with occasional stirring, and the values of the e.m.f. are recorded. A sudden fall in the negative potential of the silver electrode takes place at the end-point. The titration curves are symmetrical as the reaction involves two monovalent ions, and the end-point is taken as the point of inflexion. When a curve is not plotted the values of $\dfrac{\Delta E}{\Delta C}$ are calculated and the end-point is given by a maximum.

Percentage of mercaptan sulphur

$$= \frac{\text{ml. of AgNO}_3 \times \text{normality of AgNO}_3 \times 0.03206 \times 100}{\text{weight of mercaptan}}.$$

Accuracy $\pm 1\%$.

PREPARATIONS: *Alcoholic sodium acetate.* Sodium acetate ($CH_3.COONa, 3H_2O$) is dissolved in ethyl alcohol (96 %) and an approximately 0·1N solution is made up with ethyl alcohol.

Silver nitrate. The alcohol used in the preparation must be free from aldehydes, and therefore *iso*propyl alcohol is preferable to ethyl alcohol which latter oxidizes slowly to acetaldehyde. The *iso*propyl alcohol is freed from aldehydes by dissolving silver nitrate (0·5 g.) in the commercial alcohol (1 l.) and exposing the mixture in a clear glass bottle to sunlight for several hours. The liquid is decanted from the precipitated silver, the excess silver nitrate is removed by titration with an aqueous solution of sodium chloride and the alcohol is distilled. The azeotrope (containing 9 % of water) is used for making up a stock solution of silver nitrate (0·1 N) which can be kept in a dark bottle without deterioration for several months. The standard solution used in the titration (0·01 N) is made up by diluting the stock solution with isopropyl alcohol immediately before use.

The mercaptan solution. Sulphur and hydrogen sulphide are removed by shaking the solution with mercury and acidified cadmium sulphate (10 % $CdSO_4$, 2 % H_2SO_4).

Silver electrode. It is advisable to clean the silver electrode with a solution of potassium cyanide and wash it carefully with water immediately before the titration.

2. Potentiometric titration in aqueous-alkaline solution

The apparatus is as shown in fig. 21. The silver half-cell is an electrode of silver wire (about 2 mm. in diameter see above) immersed in an aqueous-alkaline solution. The electrolyte bridge and mercury half-cell are filled with an aqueous or an alcoholic solution of sodium acetate (0·1 N). The remainder of the apparatus is as described above. The cell is

⁻Ag	Aq. or alcoholic sodium acetate 0·1 N	Aq. or alcoholic sodium acetate 0·1 N	Hg⁺.

The potential of the silver electrode is not constant but varies with the mercaptan and the composition of the solution in the titration cell.

An alkaline solution (approximately normal) is prepared by adding water, or sodium hydroxide, to the mercaptan dissolved in sodium hydroxide, or water. Ammonia ($d = 0·880$) is added to prevent the precipitation of silver oxide during the titration (a 0·05 N solution based on ammonia is necessary). The final volume in the titration cell should be 50–100 ml. The silver electrode is placed in the cell, the solution is stirred, a standard solution of silver nitrate (0·1 N, 0·01 N or 0·001 N depending upon the concentration of the mercaptan) is added in small amounts and the values of the e.m.f. are recorded. A sudden drop in the potential of the silver electrode takes place, and the most positive potential in the almost linear descending part of the curve is taken as the end-point; this was derived empirically from the titration of a large number of standard solutions of different mercaptans.

Accuracy $\pm 1 \%$.

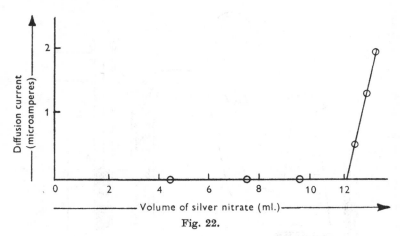

Fig. 22.

3. *Amperometric titration*

The two methods described above follow the titration by the change of potential of the indicator electrode. The end-point can be found also by measuring the diffusion current of the mercaptan, or silver nitrate, at the potential of a rotating platinum electrode.[1] The current, as measured by a micro-ammeter, during the titration is plotted against the volume of silver nitrate added, and the point of intersection of the two straight lines so obtained is the end-point (fig. 22). There is only a very small current, or none at all, as long as some mercaptan is present, but as soon as the silver nitrate is in excess the current increases rapidly.

The method has been used for the estimation of small amounts of primary, secondary and tertiary mercaptans

---

[1] I. M. Kolthoff and W. E. Harris, *loc. cit.* (p. 110, n. 2).

(0·2 mg. in 100 ml.), and in the absence of halogen ions the titration can be done in acid or neutral medium using a calomel cell as reference. When chloride or bromide ions are present ammonia is added to the cell before the titration, but then the reference electrode must be changed from the calomel cell to one slightly more negative (see below). This procedure can be used when large amounts of the chloride and small amounts of bromide ion are in solution, but large concentrations of the

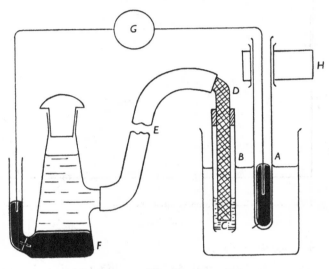

Fig. 23.

bromide ion as well as cyanide and other ions which give insoluble silver salts interfere (see p. 110).

The apparatus is as shown in fig. 23. The cell is

Pt⁻	Alcoholic solution of ammonium nitrate	Aq. potassium chloride	Potassium iodide and mercuric iodide dissolved in aq. saturated potassium chloride	Hg⁺

The mercaptan (containing about 5 mg. of sulphur as mercaptan) is weighed into a beaker (250 ml.) and dissolved in ethyl alcohol (to bring the volume up to 100 ml., 95 %). Ammonia (to give an 0·25 M solution) and some non-interfering electrolyte such as ammonium nitrate (to give a 0·01–0·1 M solution) are added, and the mixture is titrated with aqueous silver nitrate (0·005M). Two or three readings of the current are made before and after the end-point and plotted against the volume of silver nitrate added. The platinum electrode may become insensitive or erratic after it has been used for a long time or when large amounts of mercaptan are titrated. The sensitivity can be restored by wiping the electrode with a cloth.

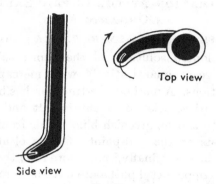

Top view

Side view

Fig. 24.

Accuracy ± 0·3 % (for concentrations of about 2 mg. of sulphur/100 ml.) ± 1–2 % (for concentrations of about 0·2 mg. of sulphur/100 ml.)

Apparatus. The platinum electrode is about 6–8 mm. long, about 0·5 mm. in diameter and sealed in soft glass tubing 6 mm. in diameter (see fig. 24). The electrode should be wiped clean with a cloth between titrations. An ordinary electric motor, *H*, is suitable for driving the electrode.

The reference electrode gives a potential of − 0·23 volt against a standard calomel cell. The electrolyte is prepared by dissolving potassium iodide (4·2 g.) and mercuric iodide (1·3 g.) in a saturated solution of potassium chloride (100 ml.). The electrode is a layer of mercury, *F*, at the bottom of the cell (see fig. 23).

The salt bridge is made from 60 cm. of 6 mm. soft rubber tubing, *E*, filled with a saturated solution of potassium chloride. The rubber is fastened to a short length of glass tubing, *D*, filled with a gel of agar (3 %) and potassium chloride (30 %). A coarsely sintered glass disk, *C*, is fitted at the end. The solution is protected from contamination with iodide by fixing a guard tube, B, with a finely sintered plug at the end. It is essential to remove all sources of high resistance from the bridge (e.g. bubbles of air).

A micro-ammeter, or a galvanometer (sensitivity of $0\cdot25\mu A$. per division), is used to measure the current.

ESTIMATION BY THE CUPRIC SALT OF AN ORGANIC ACID

$$4R.SH + 2Cu(OAc)_2 = 2CuSR + R.S.S.R + 4HOAc.$$

The highly coloured solutions of the cupric salts of certain organic acids react quantitatively with mercaptans to give colourless solutions. A method of estimation has been described using a standard solution of a cupric salt and titrating the mercaptan until a pale, greenish blue colour from an excess of the salt persists at the end-point. Cupric oleate and naphtheneate were used originally,[1] and more recently copper butyl phthalate and copper octyl phthalate have been recommended.[2] It is claimed that the second pair of reagents are of definite composition and can be prepared more easily than the first two. The procedure and the degree of accuracy is the same for all the reagents. In the estimation of mercaptans dissolved in a hydrocarbon it is advantageous to use copper octyl phthalate, as this reagent is also very soluble in hydrocarbons.

The reaction takes place rapidly and completely in a homogeneous solution and there is no interference from unsaturated compounds, organic disulphides, thiocyanates, thiocyanoacetates and hydrogen cyanide (cf. pp. 110–11). A wide variety of primary, secondary and tertiary mercaptans have been estimated successfully, but the method cannot be used for the estimation of thioglycollic acid and dithioethylene glycol. Thioglycollic acid reacts normally to give the cuprous mercaptide which then reacts with an excess of the cupric ion to form a dark blue, insoluble cupric cuprous mercaptide, but no sharp colour change takes place. Dithioethylene glycol yields a stable dark blue cupric salt on the addition of copper alkyl phthalates, but

[1] G. R. Bond, Jr., *Ind. Eng. Chem.* (Anal. ed.), 1933, **5**, 257–60.
[2] E. Turk and E. E. Reid, *Anal. Chem.* 1945, **17**, 713–14.

reduction to the cuprous state does not take place. Methyl and butyl thioglycollates however, behave normally.

The reagents are standardized by hydrolysing a known volume of the solution with an excess of dilute acid and titrating the liberated cupric ion iodometrically.

A solution of the mercaptan (containing 0·1–0·3 g. according to the molecular weight) in amyl alcohol, petrol or benzene (50 ml.) is placed in a glass-stoppered bottle and titrated slowly with a standardized solution of the copper salt. Near the end-point the mixture should be shaken vigorously after each addition (about 0·5 ml.). The solution may darken during the titration but it clears as the end-point is reached. With some mercaptans, especially those of low molecular weight, the yellow cuprous mercaptide is precipitated. The greenish blue colour of the end-point is seen easily even when this happens. An appreciable amount of the copper salt is needed to give a definite colour to 50 ml. of solution, so that a blank determination must be done under the same conditions as the estimation.

Percentage of mercaptan sulphur

$$= \frac{(\text{ml. of Cu salt} - \text{blank}) \times \text{g. Cu/ml.} \times 100 \times 1\cdot009}{\text{weight of sample}}.$$

Accuracy ± 1·5 %.

PREPARATIONS: *Cupric butyl, or octyl, phthalate solutions.* Cupric butyl phthalate (25·29 g., see below) or cupric octyl phthalate (30·9 g., see below) is weighed into a graduated flask (1 l.), dissolved in glacial acetic acid (50 ml.) and the solution made up with amyl alcohol (to 1 l.).

Copper butyl phthalate. A mixture of finely powdered phthalic anhydride (74 g., 0·5 mol.) and n-butyl alcohol (50 ml., 1·1 mol.) is heated under a reflux condenser with shaking. When the temperature reaches 105° C. heating is stopped, but shaking is continued, the temperature rises to about 120° C. as an exothermic reaction takes place. The clear solution which results after a few minutes is allowed to cool and then poured into a solution of sodium hydroxide (20 g. in 1500 ml. of water). If the solution is alkaline it is acidified (to litmus) with acetic acid. The solution is filtered and copper sulphate (65 g., $CuSO_4$, $5H_2O$,

1·04 mol.) dissolved in water (500 ml.) is added slowly with vigorous stirring. The precipitated cupric butyl phthalate is filtered off, washed thoroughly with water, dried in air, powdered and dried in a vacuum desiccator. Yield, 95 %.

Copper octyl phthalate. Octyl alcohol (68 g.) is used instead of *n*-butyl alcohol (50 ml.) in the above preparation. Copper octyl phthalate is more difficult to filter and dry than copper butyl phthalate.

Copper oleate. A warm dilute solution of elaine oil is neutralized exactly with dilute sodium hydroxide, and an aqueous solution of cupric nitrate (the theoretical amount) is then added. The solid is filtered off, washed thoroughly with water and dried as above.

Standardization of the cupric salt solutions. A measured volume of the solution (containing 50–100 mg. of copper) is hydrolysed by boiling with hydrochloric acid (10 ml., 1:1) for 2–5 min. The mixture is cooled, poured into a separating funnel, the aqueous layer is run off, the organic layer is washed with water (3 times, 15 ml.) and the washings combined with the aqueous layer. If an emulsion forms when the organic layer is washed a little isopropyl alcohol (3–5 ml.) can be added. The solution is made just alkaline with ammonia and then very slightly acid with glacial acetic acid. Potassium iodide (about 2 g.) is added and the liberated iodine is titrated with a solution of sodium thiosulphate (0·1 N),[1] which has been standardized against copper as the iodine formed may not be in the stoichiometric proportion of the following equation:

$$2CuSO_4 + 4KI = Cu_2I_2 + I_2 + 2K_2SO_4.$$

The temperature at which the copper butyl, or octyl phthalate solutions are standardized should be the same as in the estimation, as a difference of 10° C. causes an error of approximately 1 %.

Normality of copper salt

$$= \frac{\text{ml. of sodium thiosulphate} \times N \times 2}{\text{ml. of copper solution}}.$$

[1] A. S. Weatherburn, M. W. Weatherburn and C. H. Bayley, *Ind. Eng. Chem.* (Anal. ed.), 1944, **16**, 703–4.

If the solution is required to be exactly 0·1 N and if P is the purity of the cupric butyl, or octyl phthalate, then the weights of salt to be used in the preparation of the solution are respectively

$$\frac{25 \cdot 29 \times 100}{P} \quad \text{and} \quad \frac{30 \cdot 9 \times 100}{P}.$$

The solutions of the cupric alkyl phthalates should be stored in dark bottles to slow down the formation of peroxides. Solutions containing large amounts of precipitate should be thrown away.

INDIRECT ACID TITRATION

$$2R.SH + Pb(OH)^{+}.CH_{3}.COO^{-} = \begin{matrix} R.S \\ \diagdown \\ R.S \diagup \end{matrix} Pb + H_{2}O + CH_{3}.COOH,$$

$$\begin{matrix} R.S \\ \diagdown \\ R.S \diagup \end{matrix} Pb + H_{2}SO_{4} = 2R.SH + PbSO_{4}.$$

A rapid acidimetric method of analysis has been described for mercaptans and can be used when thioethers and disulphides are present.[1] Sulphur, however, interferes and must be removed by shaking the solution with mercury. A solution of the lead salt in benzene is prepared by shaking an aqueous suspension of basic lead acetate with a benzene solution of the mercaptan. The two layers are separated, and the mercaptide is decomposed by shaking the upper layer with an excess of a standard solution of acid and the excess is determined by titration with alkali.

Aqueous solutions of sodium plumbite and lead acetate have been used and found to be unsatisfactory for the preparation of the mercaptides, and solvent naphtha which has been tried as a solvent cannot replace benzene owing to the precipitation of some of the less soluble mercaptides.

A suspension of basic lead acetate (100 ml., see below) is added to a separating funnel containing a solution of the mercaptan (5–10 g.) in benzene (200–300 ml., see below). The mixture is shaken vigorously, and the benzene layer filtered into a separating funnel. Sulphuric acid (about 25 ml., excess, 0·1 N) is added and the mixture shaken vigorously until the yellow or greenish yellow colour of the mercaptide disappears

[1] W. F. Faragher, J. C. Morrell and G. S. Monroe, *J. Ind. Eng. Chem.* 1927, **19**, 1281–4.

from the benzene layer. The acid is filtered free from lead sulphate, the solid and benzene layer are washed with water and the combined washings and filtrate are titrated with a standard solution of sodium hydroxide ($0 \cdot 1 \text{N}$).

The lead sulphate can be determined gravimetrically but this takes longer and no advantage is gained.

Accuracy $\pm 2 \%$.

PREPARATIONS: *Basic lead acetate.* A suspension of basic lead acetate is prepared by dissolving lead acetate ($3 \cdot 79$ g., $Pb(O.OC.CH_3)_2, 3H_2O$) and sodium hydroxide ($0 \cdot 4$ g.) in water (100 ml.). The solid formed is not completely soluble in the water.

Benzene. The benzene used in the estimation must be pure, or freed from sulphur by shaking with mercury.

OXIDATION WITH IODINE

$$2R.SH + I_2 = R.S.S.R + 2HI.$$

The mild oxidation of a mercaptan to a disulphide can be carried out quantitatively with iodine and the reaction is used as a means of estimation. Accurate values can be obtained with aromatic mercaptans by shaking with an excess of iodine and determining the excess by titration with sodium thiosulphate.[1] The more volatile aliphatic mercaptans give low results unless precautions are taken during weighing to avoid loss by volatilization.[2] The reaction does not go to completion if the concentration of hydriodic acid is high, and this may account for some of the apparently anomalous results which have been recorded. The method is unsatisfactory when impurities which react with iodine are present. In the titration of a benzene solution of mercaptans high values may result if the apparent end-point is taken, as the blue colour reappears on standing (*n*-butyl mercaptan, $105 \cdot 8 \%$). An approximate analysis can be made by adding sodium thiosulphate in small amounts to the benzene solution and iodine and shaking vigorously until the blue colour reappears; shaking during 20 min. is usually necessary. A true

[1] P. Klason and T. Carlson, *Ber.* 1906, **39**, 738.
[2] J. W. Kimball, R. L. Kramer and E. E. Reid, *J. Amer. Chem. Soc.* 1921, **43**, 1199.

value can be obtained after allowing the mixture of the mercaptan solution and iodine to stand for 24 hr.[1]

The mercaptan (0·25 g.) in a small weighing bottle (fitted with
a well-ground glass stopper) is placed in a glass-stoppered bottle
(250 ml.) containing an aqueous potassium iodide-iodine solution (35 ml., 0·1N). The stopper of the weighing bottle is
loosened immediately before dropping into the reaction bottle.
The mercaptan and iodine are mixed thoroughly by shaking
vigorously, and the excess iodine is titrated with a standard
solution of sodium thiosulphate (0·1N) using starch (1 ml., 1 %)
as the indicator. The potassium iodide-iodine solution (35 ml.)
is titrated with the standard solution of sodium thiosulphate.

Accurate values are obtained with methyl mercaptan only
when the specimen is sealed in a bulb before weighing.

Percentage of mercaptan

$$= \frac{\text{ml. of } Na_2S_2O_3 \ (35 - \text{experiment}) \times 33 \times 100}{\text{weight of mercaptan (in grams)} \times 1000}.$$

Accuracy ± 1 %.

An alternative procedure has been tried for a small number
of mercaptans, usually dissolved in benzene, which involves
an acidimetric titration of the hydriodic acid formed in the
oxidation with iodine.[1] The method cannot be used when unsaturated compounds are present (see p. 118).

A benzene solution of the mercaptan (50 ml., containing
approximately 0·05 % sulphur) and an excess of iodine in potassium iodide are mixed in a glass-stoppered bottle and allowed to
stand until oxidation is complete (see above). The excess iodine
is removed with sodium thiosulphate solution, the aqueous and
benzene layers are separated and the latter washed with water
(three lots of 60 ml. each). The combined washings and the
aqueous layer are titrated with a standard solution of alkali
(0·025N) using bromocresol-purple or purified litmus as the
indicator.

The method is slightly less accurate than the direct titration
with iodine.

Accuracy ± 2 %.

[1] J. R. Sampey and E. E. Reid, *J. Amer. Chem. Soc.* 1932, **54**, 3404–9.

COLORIMETRIC

Folin's reagent, originally used for the estimation of cystine, was prepared from sodium tungstate and phosphoric acid, and it has been shown to be essentially phospho-18-tungstic acid (cf. pp. 102–3).[1] A rapid colorimetric method of estimation for thiol compounds has been developed using this acid, and reliable results are obtained when the test solutions are made up under carefully controlled conditions.[2] Phospho-18-tungstic acid is reduced to a blue complex by thiol compounds and many other substances, but by lowering the pH of the solution to 5 and limiting the time interval between the preparation of the solution and the colorimetric estimation the effects from the majority of the weaker reducing agents can be ignored. Strong reducing agents, however, produce the complex at a rate comparable with thiol compounds and interfere seriously with the estimation.

The reaction appears to take place in two stages, the first of which is slow and the second fast. The maximum colour is directly proportional to the concentration of the reductant, and at room temperature it is stable at pH 5 for at least 6 hr. The intensity of the colour above pH 8 is greater than that developed in more strongly acid solutions, and below pH 4·7 the colour forms slowly and the maximum intensity is less than that produced over the pH range 4·7–7·5. The colour only remains constant in this last range and an excess of the reagent must always be present. The colour can be inhibited greatly by formaldehyde, acetone, and in certain estimations (e.g. cystine) by fairly high concentrations of reagents such as cadmium chloride.

An equation suggested for the reaction is

$$2R.SH + (H_2O)_3.P_2O_5(WO_3)_{18} = R.S.S.R + H_2O + (H_2O)_3.P_2O_5(WO_3)_{17}WO_2$$
$$\text{or } (H_2O)_3.P_2O_5(WO_3)_{16}W_2O_5,$$

[1] O. Folin and W. Denis, *J. Biol. Chem.* 1912, **12**, 239–43; cf. J. W. H. Lugg, *Biochem. J.* 1932, **26**, 2144–59.

[2] K. Shinohara, *J. Biol. Chem.* 1937, **120**, 743–9; *ibid.* 1936, **112**, 671–82 and 683–96; *ibid.* 1935, **110**, 263–77; K. Shinohara and M. Kilpatrick, *ibid.* 1934, **105**, 241–51; A. Schöberl, *Ber.* 1937, **70**, 1186–93; A. Schöberl and E. Ludwig, *ibid.* 1937, **70**, 1422–32; J. W. H. Lugg, *loc. cit.* (p. 124, n. 1).

but this is not entirely satisfactory, and as an alternative it is suggested that the following reaction may take place at pH 5:

$$2R.\text{SH} + 2\text{H}_6\text{P}_2\text{O}_6(\text{WO}_3)_{18} = R.\text{S}.\text{S}.R + \text{H}_2\text{O} + [\text{H}_6\text{P}_2\text{O}_6(\text{WO}_3)_{17}]_2\text{W}_2\text{O}_5.$$

Phospho-18-tungstic acid, or the ammonium salt (4 ml.), is added to the thiol compound (1 ml.) in dilute hydrochloric acid (0·2N) and an acetate buffer solution (3 ml.; 2M acetic acid; 10 ml., 2M sodium acetate) and the mixture left to stand for 30 min. The coloured solution is diluted to 50 ml. and compared with standard solutions in a colorimeter.

Accuracy ± 5 %.

PREPARATIONS: *Phospho-18-tungstic acid solution*. See O. Folin and A. D. Marenzi, *J. Biol. Chem.* 1929, **83**, 111; cf. K. Shinohara, *ibid.* 1937, **120**, 743–9; cf. p. 102.

Ammonium phospho-18-tungstate solution. See K. Shinohara, *ibid.* 1937, **120**, 743–9.

MISCELLANEOUS

(1) The acidimetric titration of the hydrochloric acid formed in the reaction

$$R.\text{SH} + \text{HgCl}_2 = R.\text{S}.\text{HgCl} + \text{HCl}$$

has been used for the analysis of a small number of mercaptans.[1] The method often is unsatisfactory as mercuric chloride also forms hydrogen chloride with unsaturated compounds.

A benzene solution of the mercaptan (50 ml., containing about 0·05 % sulphur) and an aqueous solution of mercuric chloride (25 ml., 1 %) are shaken vigorously for 3 min. A solid may be formed, but the amount which is precipitated depends on the individual mercaptan. If no solid separates, or if only a slight precipitate is formed, the two layers are separated and the hydrochloric acid is extracted from the benzene layer with water (three lots of 25 ml., each). The combined washings and the aqueous layer are titrated with a standard solution of sodium hydroxide (0·025N) using methyl orange or methyl red as the indicator. When a heavy precipitate is formed no separation is attempted and the acid is estimated by adding the

[1] J. R. Sampey and E. E. Reid, *J. Amer. Chem. Soc.* 1932, **54**, 3404–9.

alkali slowly to the mixture with frequent shaking. Methyl orange must be used as the indicator as methyl red dissolves in the benzene layer giving an orange colour, even when the aqueous layer is acid. The end-point is determined easily. Accuracy $\pm 1\%$.

(2) Mercaptans and hydrogen sulphide have been estimated iodometrically by dissolving them in a solution of cadmium chloride, adding an excess of a standard solution of iodine and titrating this excess with sodium thiosulphate. The method has been used for the rapid determination of mercaptans in gases and aqueous solutions.[1]

(3) The active hydrogen of mercaptans and thiophenols can be estimated sometimes by the reaction with methyl magnesium iodide (Zerewitinoff's method, see pp. 55–64) or with lithium aluminium hydride (see pp. 64–6).

(4) Mercaptans can be removed completely from solution by shaking with an alcoholic plumbite solution. Quantitative analysis is possible therefore by finding the sulphur content of the solution before and after filtering off the precipitated lead salt. Sulphur can be determined by the lamp method (see para. (5) below).[2]

A mixture of the solution containing the mercaptan (2 parts) and an alcoholic plumbite solution (1 part) is shaken until no discoloration occurs when a portion is shaken with a little sulphur and aqueous sodium plumbite. The mercaptan is precipitated as its lead salt, the solid is filtered off, washed with water and the combined washings and filtrate used for the determination of sulphur.

PREPARATION: *Alcoholic plumbite solution.* Equal volumes are mixed of alcohol (95 %) and a solution prepared by saturating aqueous sodium hydroxide (20–25 %) with litharge and filtering off the excess solid.

(5) The total sulphur content of a solution (usually in petroleum or a similar solvent) can be found by burning in

---

[1] J. A. Shaw, *Ind. Eng. Chem.* (Anal. ed.), 1940, 12, 668–71.
[2] W. F. Faragher, J. C. Morrell and G. S. Monroe, *J. Ind. Eng. Chem.* 1927, 19, 1281–4; G. L. Wendt and S. H. Diggs, *ibid.* 1924, 16, 1113–15; cf. M. A. Youtz and P. P. Perkins, *ibid.* 1927, 19, 1250.

a specially designed wick lamp, absorbing the gases in a standard solution of sodium carbonate and titrating the excess alkali with a standard solution of hydrochloric acid.[1]

(6) An automatic apparatus has been developed for recording small concentrations of sulphur dioxide in the air, and its use can be extended to other classes of compounds including mercaptans, thiophenols, and halogenated compounds such as chloroform, carbon tetrachloride and ethylene dichloride.[2]

(7) An approximate estimation of the mercaptan content of petroleum samples can be made by titration with a standard solution of ammoniacal cuprous chloride after the specimen has been freed from hydrogen sulphide by washing with a dilute solution of sodium carbonate.[3]

(8) Mercaptans can be estimated by shaking with flowers of sulphur and sodium plumbite in sodium hydroxide. The lead sulphide formed is oxidized to and estimated gravimetrically as lead sulphate.[4]

(9) Mercaptans (in petrol) may be estimated by direct titration with silver nitrate in the presence of alkali (such as sodium cresylate) and an amine (such as pyridine) using sodium nitroprusside as an internal indicator.[5] A homogeneous reaction mixture is obtained by adding either butyl or amyl alcohol.

(10) The following colorimetric methods have been used:

(a) A solution of zinc acetate (2 ml., $0\cdot2$M) and ammonium hydroxide (2 ml., 1M) are added to a solution of the mercaptan (5 ml.) which has been cooled to 25° C. Sodium nitroprusside ($0\cdot5$ ml., 5%) is then added and the mixture shaken vigorously. Standard solutions are used for comparison.

It is claimed that the method is specific for water-soluble thiol compounds, but although it is simple and rapid it lacks

[1] *Standard Methods for Testing Petroleum and its Products*, 7th ed. 1946, pp. 385–9; cf. pp. 390–406; W. H. Lane, *Anal. Chem.* 1948, **20**, 1045–7.
[2] M. D. Thomas, J. O. Ivie, J. N. Abersold and R. H. Hendricks, *Ind. Eng. Chem.* (Anal. ed.), 1943, **15**, 287–90.
[3] W. Krause, *Chemist-Analyst*, 1938, no. 1, **27**, 14.
[4] G. L. Wendt and S. H. Diggs, *J. Ind. Eng. Chem.* 1924, **16**, 1113–15.
[5] G. E. Mapstone, *Australian Chem. Inst. J. and Proc.* 1948, **15**, 236–41.

precision and therefore is only useful for rough estimations.[1] The minimum concentration of cysteine which can be estimated by this method is 4×10^{-5}M.

(b) Cobaltous chloride solution (2 ml., 0·2M) is added to a solution of the mercaptan (5 ml.) which has been cooled to 25° C. The mixture is shaken, made strongly alkaline with sodium hydroxide (4 ml., 0·5N) and then shaken again. Hydrogen peroxide (3 %) is added, the cobaltic hydroxide is removed either by filtration or by centrifuging, and the brownish yellow solution is compared with standards.

This method has been used frequently, and many modifications have been proposed to overcome difficulties associated with non-specificity and impermanency of the colour.[2]

(c) Mercaptans react with dimethyl-p-phenylenediamine in the presence of hydrogen ions and ferric alum to give a blue coloration which is proportional to the concentration of the mercaptan. Sulphides interfere with the estimation as they undergo the same reaction.[3]

(d) Mercaptans in gases may be estimated by their reaction with copper butyl phthalate.[4] A known volume of the gas is passed through a dry bead-type absorber containing a solution of copper butyl phthalate in glacial acetic acid and butyl alcohol. The solution which results is compared with standards.

[1] K. Shinohara and M. Kilpatrick, *J. Biol. Chem.* 1934, **105**, 241–51; F. S. Hammett and S. S. Chapman, *J. Lab. Chim Med.* 1938, **24**, 293; cf. V. Arnold, *Z. physiol. Chem.* 1911, **70**, 300.

[2] K. Shinohara and M. Kilpatrick, *loc. cit.* (p. 128, n. 1); cf. L. Michaelis and co-workers, *ibid.* 1929, **83**, 191–210 and 367–73; *J. Amer. Chem. Soc.* 1930, **52**, 4418; E. C. Kendall and J. E. Holst, *J. Biol. Chem.* 1931, **91**, 435–74.

[3] H. Toyoda, *Bull. Chem. Soc. Japan*, 1934, **9**, 263; cf. K. Shinohara, *J. Biol. Chem.* 1935, **109**, 665–79.

[4] D. L. White and F. E. Reichardt, *Gas*, 1949, **25**, no. 6, 38–9; cf. G. R. Bond, Jr., *Ind. Eng. Chem.* (Anal. ed.), 1933, **5**, 257–60 and p. 118.

CHAPTER IV

CARBONYL COMPOUNDS

ALDEHYDES AND KETONES

The general reactions for the estimation of aldehydes and ketones follow closely the reactions used for their characterization. Condensation reactions are used followed by analysis of one of the products of the reaction or the excess of the reagent used. The kinetics and mechanism of typical condensations with semicarbazide have been examined closely,[1] and although reactions using other reagents have not been investigated as fully, it seems probable that the same conditions apply. The condensations are balanced reactions so that it is impossible theoretically to make them go to completion, but experimental conditions can be found which fulfil this condition for practical purposes.

The effect of configuration upon the rate of formation of condensation products has been investigated frequently and certain qualitative facts have been established. Addition and condensation reactions of carbonyl compounds are faster with aldehydes than with ketones, and an increase in the size of the groups attached to the carbonyl group retards the reaction. In the aromatic series the introduction of substituents into the *meta-* and *para*-positions of the carbonyl compound has very little effect upon the rate of formation of condensation products, but the presence of groups in one or both of the *ortho*-positions retards the reaction seriously, regardless of the nature of the substituents. Thus, acetophenone forms the normal condensation products readily, but the introduction of a methyl group into both of the *ortho*-positions suppresses these reactions completely. This general phenomenon was attributed originally to 'steric hindrance', but evidence has accumulated which

[1] J. B. Conant and P. D. Bartlett, *J. Amer. Chem. Soc.* 1932, **54**, 2881-99; F. H. Westheimer, *ibid.* 1934, **56**, 1962-5.

shows that the loss of reactivity is not dependent entirely upon the bulk of the *ortho*-substituents as implied by the above term, and the condition is more conveniently termed the '*ortho* effect' or 'proximity effect'.[1]

Special methods of estimation for particular aldehydes and ketones also are described and depend on characteristic reactions of the compounds themselves and not entirely on those of the carbonyl grouping.

The following methods of estimation are available for the quantitative analysis of aldehydes and ketones:

Preparation of the bisulphite compound and estimation of the excess bisulphite:

A. Titration with iodine. Ripper's method:

$$\underset{R'}{\overset{R}{\diagdown}}C{=}O + NaHSO_3 \rightleftharpoons \underset{R'}{\overset{R}{\diagdown}}C{\overset{OH}{\underset{SO_3Na}{\diagup}}},$$

$$H_2SO_3 + H_2O + I_2 = 2HI + H_2SO_4.$$

B. Titration with sodium hydroxide:

$$R.CHO + Na_2SO_3 + H_2O \rightleftharpoons R.\underset{SO_3Na}{\overset{OH}{\underset{|}{\overset{|}{C}}}}H + NaOH.$$

Preparation of phenylhydrazones and estimation of the excess phenylhydrazine:

$$\underset{R'}{\overset{R}{\diagdown}}C{=}O + H_2N.NH.C_6H_5 \rightleftharpoons \underset{R'}{\overset{R}{\diagdown}}C{=}N.NH.C_6H_5 + H_2O.$$

A. Oxidation by Fehling's solution:

$$2C_6H_5.NH.NH_2 + 3O = C_6H_6 + 2N_2 + C_6H_5OH + 2H_2O.$$

B. Oxidation by iodine:

$$C_6H_5.NH.NH_2 + 2I_2 = 3HI + N_2 + C_6H_5I.$$

Preparation and gravimetric estimation of 2:4-dinitrophenylhydrazones:

$$\underset{R'}{\overset{R}{\diagdown}}C{=}O + H_2N.NH{-}\hspace{-2pt}\underset{NO_2}{\diagdown}\hspace{-2pt}{-}NO_2 = \underset{R'}{\overset{R}{\diagdown}}C{=}N.NH{-}\hspace{-2pt}\underset{NO_2}{\diagdown}\hspace{-2pt}{-}NO_2 + H_2O.$$

[1] J. F. J. Dippy and R. H. Lewis, *J. Chem. Soc.* 1937, p. 1426; J. F. J. Dippy, D. P. Evans, R. H. Lewis and H. B. Watson, *ibid.* 1937, p. 1421; cf. V. Meyer, *Ber.* 1894, **27**, 510–14.

Preparation of the oxime and estimation of either the hydrochloric acid or the water formed in the reaction:

$$\underset{R'}{\overset{R}{>}}C{=}O + H_2NOH.HCl \rightleftharpoons \underset{R'}{\overset{R}{>}}C{=}NOH + H_2O + HCl.$$

A. Methods for the estimation of the hydrochloric acid.

B. Estimation of the water.

Miscellaneous.

PREPARATION OF THE BISULPHITE COMPOUND AND ESTIMATION OF THE EXCESS BISULPHITE

A. Titration with Iodine. Ripper's Method

$$\underset{R'}{\overset{R}{>}}C{=}O + NaHSO_3 \rightleftharpoons \underset{R'}{\overset{R}{>}}C\underset{SO_3Na}{\overset{OH}{<}}.$$

The estimation of a bisulphite by titration with iodine is the method used most frequently to determine the percentage of the carbonyl group in a compound.[1] An excess of sodium, or potassium bisulphite, is added to the aldehyde or ketone, and after the addition compound has been formed the residual bisulphite is estimated iodometrically:

$$H_2SO_3 + H_2O + I_2 = 2HI + H_2SO_4.$$

The formation of an addition product from a carbonyl compound and a bisulphite is a reversible reaction, and the distribution at equilibrium depends upon the aldehyde or ketone, the pH of the reaction mixture, the temperature and concentration of the solution, and the excess of bisulphite. The results of analysis depend further on the specific rates of the addition and dissociation reactions which are also affected by the above conditions. The equilibrium conditions for the formation of bisulphite addition compounds have been examined,[2] and for benzaldehyde a more detailed study has been made.[3]

[1] A. E. Parkinson and E. C. Wagner, *Ind. Eng. Chem.* (Anal. ed.), 1934, **6**, 433–6; cf. A. Jolles, *Z. anal. Chem.* 1907, **46**, 764–71; *Ber.* 1906, **39**, 1306–7; B. G. Feinberg, *Am. Chem. J.* 1913, **49**, 87–116; M. Ripper, *Monatsh.* 1900, **21**, 1079–85.

[2] W. Kerp, *Arb. kais. Ges.-A.* 1904, **21**, 156–79; W. Kerp and E. Baur, *ibid.* 1907, **26**, 231–68 and 269–96; W. Kerp and P. Wöhler, *ibid.* 1909, **32**, 120.

[3] T. D. Stewart and L. H. Donnally, *J. Amer. Chem. Soc.* 1932, **54**, 2333–40, 3555–8 and 3559–69.

From Kerp's data it has been possible to calculate the error in the analysis inherent from dissociation.[1] The evidence available allows the following conclusions to be drawn about the estimation:

(a) The accuracy of the analysis depends primarily on K, the equilibrium constant for the estimation. When this is greater than 10^{-3} low results are always obtained; with a value of 10^{-3} (e.g. acetone) accurate determination is possible only by using a large excess of bisulphite; but when K is 10^{-4} or less (e.g. formaldehyde, acetaldehyde, furfuraldehyde and benzaldehyde) accurate analysis is possible. The accuracy is improved by using an excess of bisulphite and by increasing the concentration of the reagents, especially when the equilibrium constant is unfavourably high. When K is of the order of 10^{-7} or 10^{-6} (e.g. formaldehyde and acetaldehyde) even dilute solutions can be estimated using only a moderate excess of bisulphite.

(b) The accuracy can be improved by cooling the reaction mixture as this lowers the value of the equilibrium constant for the dissociation, but the time needed to reach equilibrium may exceed the 15–60 min. usually given. In the analysis of some aldehydes the results are accurate only when the titration is done at a low temperature. In the aliphatic series cooling is generally favourable although the effect is small, but with the aromatic carbonyl compounds it has a pronounced effect. The reaction mixture may be cooled immediately before and during the titration after doing the initial reaction at room temperature.

(c) The rates of the addition and dissociation reactions are affected by the hydrogen-ion concentration. The former decreases as the acidity increases, whereas the latter and the dissociation constant are at a minimum in acid solution and increase as the reaction mixture becomes less acid. A procedure has been proposed to overcome this difficulty but experimentally it has no advantage. The excess bisulphite is titrated as usual, the pH of the solution is brought to 8 by adding sodium bicarbonate and the bisulphite formed from the addition compound is then estimated by a further iodine titration.[2]

[1] I. M. Kolthoff, H. Menzel and N. H. Furman, *Volumetric Analysis*, **2**, 399.
[2] Y. Tomoda, *J. Soc. Chem. Ind.* 1929, **48**, 79 T; C. H. Lea, *Ind. Eng. Chem.* (Anal. ed.), 1934, **6**, 241–6; F. H. Goldman and H. Yagoda, *ibid.* 1943, **15**, 377–8; F. P. Clift and R. P. Cook, *Biochem. J.* 1932, **26**, 1788–99.

(d) Inaccuracies in the method frequently are due to a failure to appreciate the reactions involved and the conditions which are essential for the iodine titration. Irregular results have been attributed to reduction of the bisulphite to sulphur by the hydrogen iodide of the reaction, to the loss of sulphur dioxide by volatilization, and to air oxidation of the bisulphite. The last is probably the most serious. It is known that the accurate estimation of sulphites is only possible when the sulphite is run into the iodine and that the reverse procedure gives low results. In this determination it is impossible to run the bisulphite into the iodine, but with care reproducible and fairly accurate values can be obtained by direct titration with iodine. The estimation can be made more accurate by a back-titration. The bisulphite addition compounds do not react with iodine but reduce it only after they dissociate, so that a back-titration can be used when the rate of dissociation is not too high (e.g. formaldehyde, acetaldehyde, furfuraldehyde and benzaldehyde).

A solution of sodium, or potassium, bisulphite (25 ml., 0·3–0·4 M), is measured into two graduated flasks (50 ml.) and the carbonyl compound (0·002–0·003 mol.) is introduced into one of them (see below). The two flasks are made up (to 50 ml.) with distilled water, shaken, stoppered and left to stand for 15–60 min. Equilibrium usually is reached at the end of this time, and in many cases a shorter period is satisfactory.

Two methods of estimation are possible, a direct titration of the bisulphite with iodine and a back-titration after the addition of an excess of iodine. Agreement between the two methods is usually within 0·5 %.

(i) The two solutions (25 ml. each) are transferred to conical flasks, the tip of the pipette is placed near to the bottom of the flask and allowed to drain. The titration with iodine (0·1 N) is done immediately and rapidly using starch (5–10 ml., 0·5 %) as the indicator near the end of the titration. The end-point is usually satisfactory with aliphatic aldehydes, but with aromatic compounds it is transitory, persisting only a few seconds, and unless the titration is done rapidly it is inevitably overrun. A blank estimation should be done in parallel.

(ii) The solutions (10 ml. each) are run from pipettes into an excess of iodine (50 ml., 0·1 N) in conical flasks which are rotated continuously during the addition. The excess of iodine is titrated immediately with a solution of sodium thiosulphate (0·1 N). The end-point is sharp, but frequently it is an advantage to cool the iodine in an ice-bath immediately before the addition of the reaction mixtures.

A low-boiling aldehyde (e.g. acetaldehyde) is weighed in an ampoule which is broken under the surface of the bisulphite and allowance is made for the volume of the glass in making up the solution to a given volume. Water-insoluble compounds are added to the bisulphite on top of which a layer of aldehyde-free ethyl alcohol (5–10 ml.) is run.

B. Titration with Sodium Hydroxide

$$R.\text{CHO} + \text{Na}_2\text{SO}_3 + \text{H}_2\text{O} \rightleftharpoons R.\overset{\displaystyle \text{OH}}{\underset{\displaystyle \text{SO}_3\text{Na}}{\text{CH}}} + \text{NaOH}.$$

The bisulphite addition compound and an equivalent amount of sodium hydroxide are formed when a carbonyl compound reacts with an aqueous solution of sodium sulphite. The reaction has been used for the estimation of aldehydes and ketones by finding the increase in alkalinity of the reaction mixture.[1] The reaction is a balanced one, and many modifications of the original procedure have been tried to force it to completion and overcome the loss of low-boiling aldehydes. A mixed reagent of sodium sulphite-sodium bisulphite sometimes gives good results as the bisulphite moves the equilibrium nearer to the complete formation of the addition compound.[2] A better method which appears to overcome the instability of the bisulphite solution is to add a measured volume of a standard solution of sulphuric acid to a large excess of sodium sulphite immediately before the addition of the carbonyl compound and so produce the bisulphite *in situ*.[3] The addition compound is formed and the

[1] A. Seyewetz and J. Bardin, *Bull. soc. chim.* 1905, (3), **33**, 1000–2; cf. A. Seyewetz and Gibello, *ibid.* 1904, (3), **31**, 691–4.

[2] G. Romeo and E. D'Amico, *Ann. chim. applicata*, 1925, **15**, 320–30.

[3] S. Siggia and W. Maxcy, *Anal. Chem.* 1947, **19**, 1023–5.

excess bisulphite is titrated with a standard solution of sodium hydroxide using a pH meter to determine the end-point.

The reaction gives good results as the large excess of sodium sulphite drives the reaction almost to completion, prevents the loss of low-boiling compounds and gives a reaction mixture which dissolves many of the higher boiling insoluble aldehydes. The method is of little value for the estimation of ketones, as with the majority the titration curve of pH against the amount of alkali added rises steadily and shows no definite end-point. Acetals do not interfere as the solution is never sufficiently acid for hydrolysis to be appreciable. Acidic and basic impurities must be neutralized before the estimation is carried out.

Sulphuric acid (50 ml., 1 N) is added to an aqueous solution of sodium sulphite (250 ml., 1 M) in a glass-stoppered Erlenmeyer flask (500 ml.) which is swirled as the acid is added to prevent the loss of sulphur dioxide. The aldehyde (0·02–0·04 mol., sealed in a glass ampoule) is added, the stopper (greased for an aldehyde with a low boiling-point) is placed in position and the ampoule is broken by shaking the flask vigorously (a few glass beads may be needed in the flask). The addition compound is formed by shaking for 2–3 min. for aldehydes miscible with water and for 5 min. for the more insoluble aldehydes. The mixture is transferred completely to a beaker and stirred mechanically, the electrodes of a pH meter are placed in it and standard sodium hydroxide (0·1 N) is added drop by drop. A graph is plotted of the alkali added against the pH from which the end-point is found. A more rapid but slightly less accurate method for determining the end-point is to titrate to a pH value which has been found previously for the aldehyde. The following values have been recorded:

Aldehyde	pH range at the end-point
Acetaldehyde	9·05–9·15
Propionaldehyde	9·30–9·50
Butyraldehyde	9·40–9·50
*Crotonaldehyde	9·20–9·40
Benzaldehyde	8·85–9·05
*Cinnamaldehyde	9·50–9·60

* 2 moles of bisulphite are needed for each mol. of aldehyde.

PREPARATION: *Sodium sulphite solution.* Sodium sulphite usually contains a small amount of free alkali, and as a large volume (about 250 ml.) of the molar solution is used in each titration a significant amount of sulphuric acid (0·4–0·5 ml., 0·1 N) is needed for neutralization. It is convenient to neutralize the free acid by adding just enough of a sodium bisulphite solution (1 M) to bring the pH to 9·1.

PREPARATION OF PHENYLHYDRAZONES AND ESTIMATION OF THE EXCESS PHENYLHYDRAZINE

$$\begin{matrix} R \\ R' \end{matrix}\!\!>\!\!C{=}O + H_2N.NH.C_6H_5 \rightleftharpoons \begin{matrix} R \\ R' \end{matrix}\!\!>\!\!C{=}N.NH.C_6H_5 + H_2O.$$

The preparation of phenylhydrazones, using an excess of the reagent, followed by estimation of this excess is the basis of a number of methods which have been described for the determination of the percentage of the carbonyl group in aldehydes and ketones. Two general methods are used to estimate the excess of phenylhydrazine, one of which utilizes the products of oxidation with Fehling's solution,[1] and the other in which the reagent is titrated with iodine.[2]

A. Oxidation by Fehling's Solution

$$2C_6H_5.NH.NH_2 + 3O = C_6H_6 + 2N_2 + C_6H_5OH + 2H_2O.$$

The oxidation of phenylhydrazine with Fehling's solution goes to equilibrium within half an hour at room temperature in accordance with the above equation. The velocity of reaction increases as the temperature is raised, and at 100° C. benzene and nitrogen are formed very rapidly, but now the main reaction is

$$C_6H_5.NH.NH_2 + O = C_6H_6 + N_2 + H_2O.$$

The temperature of the reaction must be controlled therefore if the reaction is to be quantitative.

Cuprous oxide is one of the reaction products from the oxidation of phenylhydrazine with Fehling's solution, and it can be estimated by dissolving in an acid solution of ferric sulphate to

---

[1] H. Strache, *Monatsh.* 1891, **12**, 524–32; 1892, **13**, 299–315; W. Smith, *Chem. News*, 1906, **93**, 83–4; I. S. MacClean, *Biochem. J.* 1913, **7**, 611–15; G. W. Ellis, *J. Chem. Soc.* 1927, 848–51; S. Marks and R. S. Morrell, *Analyst*, 1931, **56**, 508–14.

[2] E. von Meyer, *J. prakt. Chem.* 1887, [2], **36**, 115–16; E. G. R. Ardagh and J. G. Williams, *J. Amer. Chem. Soc.* 1925, **47**, 2983–88.

give an equivalent of the ferrous salt which can be titrated with potassium permanganate.[1] The results are fairly accurate for pyruvic acid, and benzaldehyde whose hydrazones are practically insoluble in water but are unreliable for carbonyl compounds forming hydrazones which are hydrolysed in water.

A more reliable method, described originally by Strache[2] and since modified by other workers,[3] is to collect and measure the nitrogen evolved. Accurate results are obtained if precautions are taken to avoid

(i)   Incomplete formation of the hydrazone.

(ii) Hydrolysis of the hydrazone.

(iii) The hydrazone reacting with the boiling Fehling's solution.

(iv) Some of the nitrogen evolved during the oxidation of the phenylhydrazine remaining dissolved either in the reaction mixture, or in the potassium hydroxide solution below the eudiometer tube.

(v)   Atmospheric oxidation of the phenylhydrazine.

The first three sources of error all give a low result, the fourth can give a high or low result depending on whether more or less nitrogen is retained in the blank estimation, and the fifth possibility gives a high result. When the hydrazone is soluble and cannot be filtered off (e.g. acetone) hydrolysis takes place and a low value is obtained. Benzene vapour is formed as well as nitrogen in the oxidation, and rather than attempt its absorption it is convenient to place a few drops of benzene in the eudiometer tube and make allowance in the calculation for the nitrogen being saturated with benzene vapour.

An apparatus designed specially for the estimation is necessary (fig. 25) to ensure the nitrogen is evolved completely and removed rapidly from the solution by vigorous boiling of the reaction mixture under reduced pressure. This is done by connecting the reaction flask with a movable mercury reservoir.

Mercury (about 250 ml.) is placed in a strong, short-necked, round-bottomed flask (about 300 ml.) and connected with a

---

[1] I. S. MacClean, *loc. cit.* (p. 136, n. 1).

[2] H. Strache, *loc. cit.* (p. 136, n. 1).

[3] W. Smith, *loc. cit.* (p. 136, n. 1); G. W. Ellis, *loc. cit.* (p. 136, n. 1); and S. Marks and R. S. Morrell, *loc. cit.* (p. 136, n. 1).

smaller quantity in a reservoir by a glass tube (about 3 mm.
bore) and rubber tubing. Fehling's solution (about 50 ml.) is
placed above the mercury, and the flask is surrounded almost
completely by a beaker containing water. A cup, into which
a filter funnel can be fitted, and a capillary tube passing to
a eudiometer tube filled with, and standing in a trough of,
aqueous potassium hydroxide solution (30 %) complete the
connexions to the flask.

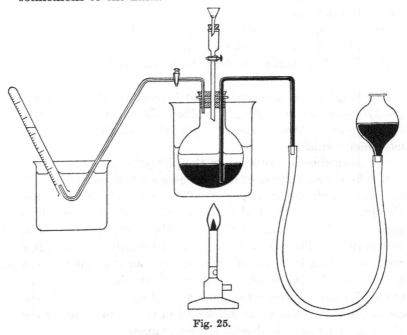

Fig. 25.

The Fehling's solution is freed first from dissolved gases by
keeping the water in the beaker boiling rapidly and by lowering
the mercury reservoir as far as possible to reduce the pressure in
the flask. The rubber tubing is pinched by a clip when most of
the mercury has been forced into the reservoir by the vapour
formed, and then a fine jet of cold water is directed on to the
upper surface of the flask. The reduced pressure in the flask
causes the Fehling's solution to boil vigorously, and by main-
taining a difference of temperature as great as possible between
the upper and lower parts of the flask the greater part of the

dissolved gas is removed from solution. After about 5 min.
constant attention to these conditions, the clip is removed, and
the gases are forced out of the apparatus. The process is
repeated until only a small break, or a slight froth, passes
through the capillary tube. The operation takes about 15 min.
if only 50 ml. of Fehling's solution are used. It is necessary to
apply the same procedure at the end of each part of the
estimation to measure all the free nitrogen from the oxidation
of the phenylhydrazine.

A mixture of phenylhydrazine hydrochloride (0·72 g.) and
sodium acetate (0·72 g.) is dissolved in water and made up in
a graduated flask of known volume (10 ml.), filtered, and placed
in a burette. Two tubes are made up, one with the reagent only
(2 ml., from the burette) and the second with the reagent (2 ml.,
from the burette) and the carbonyl compound (see below). The
tubes are heated in a boiling water-bath for a few minutes. The
contents of the tube with reagent only are placed in the cup of
the apparatus and drawn into the flask by lowering the mercury
reservoir carefully. The tube and cup are washed several times
with water and the washings drawn into the flask and brought
into the Fehling's solution. The reservoir is then lowered until
vapour is formed at the top of the flask, thereby mixing the
reagents and liberating nitrogen and benzene vapour. The
greater part of the mercury is allowed to pass into the reservoir,
and the water in the beaker is kept boiling for 3 or 4 min. until
most of the nitrogen is transferred to the eudiometer. The
remainder of the nitrogen, most of which is in solution, is removed as described above for the removal of permanent gases
in the preliminary operation.

The excess of phenylhydrazine in the second tube is then
estimated. A funnel is placed in the cup of the apparatus and
the hydrazone is filtered through glass wool, or asbestos fibre,
and the liquid drawn into the flask by lowering the mercury
reservoir. The hydrazone is washed as above so that the total
bulk of liquid is the same in the estimation of the excess as in
the blank determination.

If the carbonyl compound is insoluble in water an alcoholic,
glacial acetic acid, or aqueous acetic acid solution can be used.

When alcohol is the solvent the reservoir must be lowered carefully owing to the rapid formation of vapour which may force the Fehling's solution out of the flask. If acetic acid is used enough potassium hydroxide solution must be added to the Fehling's solution to keep it alkaline.

For estimations in which more than 0·1 g. of the substance is taken it is usually necessary to take 200 ml. of Fehling's solution, which makes the method longer, more tedious, and less accurate.

Accuracy ± 2%.

B. Oxidation by Iodine

$$C_6H_5.NH.NH_2 + 2I_2 = 3HI + N_2 + C_6H_5I.$$

The iodometric estimation of phenylhydrazine which was first investigated by von Meyer[1] has been made reliable and gives reproducible results when the conditions are carefully controlled.[2] Care must be taken to prevent oxidation of the phenylhydrazine through contact with air, the pH must be kept between 5 and 7, and the equilibrium conditions must be arranged for the hydrazone to be formed quantitatively. Air oxidation can be prevented by using boiled distilled water throughout the estimation and by keeping an inert atmosphere above the solutions. The hydrogen-ion concentration is regulated by using a standard buffer solution, usually disodium phosphate, and, although the reaction is reversible and it is impossible theoretically for the hydrazone to be formed completely, the reaction can be made to go to completion for purposes of estimation by salting out.

Accurate results are claimed for the method in a limited number of estimations, but when an aromatic nucleus is present in the carbonyl compound the rate of hydrazone formation is slow and a sufficient time must be allowed for the reaction to take place. Acetophenone needs 2 hr. instead of the 30 min. normal for aliphatic aldehydes. When two aromatic rings are present the reaction is slower still and the hydrazone is formed incompletely (e.g. only 10–15% of benzophenone phenylhydrazone is formed when equilibrium is established).

[1] E. von Meyer, loc. cit. (p. 136, n. 2).
[2] E. R. G. Ardagh and J. G. Williams, loc. cit. (p. 136, n. 2).

The phenylhydrazine is estimated by adding an excess of a standard solution of iodine which gives a mixture of iodobenzene, nitrogen and hydrogen iodide according to the above equation. An excess of sodium thiosulphate is added to the mixture, the iodine is freed from drops of iodobenzene by shaking with diethyl ether and the excess of thiosulphate is found by back-titration with a standard solution of iodine. The estimation must be done in an atmosphere of nitrogen after the addition of iodine to the phenylhydrazine in order to prevent the oxidation of the hydrogen iodide.

Solutions of phenylhydrazine hydrochloride (20 ml., 0·5M, see below), disodium phosphate (20 ml., see below) and the carbonyl compound (enough to react with about half of the phenylhydrazine, see below) are mixed in a graduated flask (75 or 100 ml.) and made up to the mark with a saturated solution of sodium chloride. The mixture is left to stand at room temperature, in the stoppered flask, until the hydrazone is formed completely. About 30 min. is usually needed for aliphatic compounds, but longer is needed for aromatic ketones (see above). If a solid separates, it should be filtered off as quickly as possible. The aqueous solution (25 ml.) is transferred to a separating funnel (filled with nitrogen, 100 ml. approximately), petrol ether (4–5 ml., 60–80° C.) is added and the mixture shaken for 2 min. After standing for 2 min. some of the aqueous layer is run into a cylinder (25 ml.), and an accurate volume (10 ml.) is transferred by a pipette to a flask fitted with a ground-glass stopper (250 ml.). Methyl orange (2 drops, 0·1 % solution) is added and the solution just acidified with dilute sulphuric or hydrochloric acid. An excess (about 5 ml.) of iodine (0·1N) is added and the mixture left to stand for 5 min. A freshly prepared solution of starch is added, and an excess (3–4 ml.) of sodium thiosulphate (0·1N) is run in, diethyl ethyl ether added (about 5 ml.) and the mixture shaken. The excess of sodium thiosulphate is determined by back-titration with iodine (0·1N).

A blank estimation should be carried out with phenylhydrazine.

Accuracy ± 0·5 %.

PREPARATIONS: *The water* used throughout the estimation should be drawn from a supply of boiled, distilled water which has been cooled and stored under an atmosphere of nitrogen.

The phenylhydrazine hydrochloride should be recrystallized before making up the solution.

Disodium phosphate. As a 0·5 M solution of disodium phosphate crystallizes out at 25° C. an equivalent amount of a slightly weaker solution may be used.

Carbonyl compound. If the carbonyl compound is liquid it is weighed out by sucking into a weighed glass bulb which is broken under water, or a mixture of water and alcohol, and the solution made up to a known volume.

PREPARATION AND GRAVIMETRIC ESTIMATION OF 2 : 4-DINITROPHENYLHYDRAZONES

$$\begin{array}{l}R \\ R'\end{array}\!\!\!>\!\!C\!=\!O + H_2N.NH\!\!\!\diagdown\!\!\!\diagup\!\!\!-NO_2 = \begin{array}{l}R \\ R'\end{array}\!\!\!>\!\!C\!=\!N.NH\!\!\!\diagdown\!\!\!\diagup\!\!\!-NO_2 + H_2O.$$
$$\qquad\qquad\qquad NO_2 \qquad\qquad\qquad\qquad\qquad NO_2$$

2:4-Dinitrophenylhydrazine has been used frequently as a reagent for the isolation, and characterization of aldehydes and ketones.[1] The derivatives are formed in good yield, and attempts have been made to find conditions by which the pure hydrazones can be prepared quantitatively so that the reagent can be used for the gravimetric estimation of carbonyl compounds (cf. p. 154). A number of methods suitable for particular carbonyl compounds[2] and some general procedures[3] have been described. The first of the general methods deals with water-soluble compounds only and describes the precipitation of the 2:4-dinitrophenyl-hydrazone from a saturated solution of the reagent at 0° C. Vanillin gives a consistently high result (102·18 %) owing to

[1] F. Wild, *Characterisation of Organic Compounds*, pp. 110–16.
[2] E. Simon, *Biochem. Z.* 1932, **247**, 171–7; H. Reynolds, O. L. Osburn and C. H. Werkman, *Iowa State Coll. J. Sci.* 1933, **7**, 443–51; O. Fernández, L. Socias and C. Torres, *Anales Soc. españ. fís. quím.* 1932, **30**, 37–49; C. Torres and S. Brosa, *ibid.* 1933, **31**, 34–6 and 649–62; F. P. Clift and R. P. Cook, *Biochem. J.* 1932, **26**, 1800–3; R. E. Houghton, *Am. J. Pharm.* 1934, **106**, 62–4; J. A. Sozzi, *Anales farm. bioquím.* (Buenos Aires), 1943, **14**, 41–7.
[3] H. A. Iddles and C. E. Jackson, *Ind. Eng. Chem.* (Anal. ed.), 1934, **6**, 454–6; H. A. Iddles, A. W. Low, B. D. Rosen and R. T. Hart, *ibid.* 1939, **11**, 102–3.

occlusion of the reagent in the crystals. It has been suggested that occlusion will always take place if the temperature of reaction is 0° C. when the saturated solution of the reagent is prepared at room temperature.[1] Identical results are obtained however at 0° C. and at room temperature, and as it is more convenient to carry out the reaction without cooling, room temperature has been adopted as the standard. The method has been extended to substances insoluble in water by dissolving them in ethyl alcohol, free from aldehydes and ketones.[2]

The determination is long and tedious and has not the accuracy normally associated with gravimetric methods, as the formation of hydrazones is a balanced reaction which makes it theoretically impossible for the derivative to be formed completely. The results of a limited number of estimations show a variation from 93·2% (mesityl oxide) to 103·5% (piperonal) with a fairly large number between 99 and 100% (p-hydroxybenzaldehyde, acetophenone, benzoin). Low yields of the derivative are obtained in the aliphatic series, especially with compounds of low molecular weight. The results in the aromatic series are usually satisfactory except when occlusion occurs. This error does not seem to be reduced by repeated washing of the precipitate or by greater care in the formation of the derivative.

p-Nitro- and p-bromo-phenylhydrazine also have been tried as reagents, but the results are not as good as with 2:4-dinitrophenylhydrazine. A lower, and less uniform yield is obtained.[3]

1. *Water-soluble aldehydes and ketones*

An aqueous solution of the carbonyl compound (a known weight) is added dropwise to an excess (50–100%) of a saturated solution of 2:4-dinitrophenylhydrazine in hydrochloric acid (2N). Hydrochloric acid (50 ml., 2N) is added, and the reaction mixture is then allowed to stand at room temperature, or in an ice-bath, for 1 hr. The solid derivative is filtered into a weighed Gooch crucible, washed with hydrochloric acid (2N), water, and dried finally to constant weight at 105–110° C.

[1] G. W. Perkins and M. W. Edwards, *Am. J. Pharm.* 1935, **107**, 208–11.
[2] H. A. Iddles, A. W. Low, B. D. Rosen and R. T. Hart, *loc. cit.* (p. 142, n. 3).
[3] B. G. Feinberg, *Am. Chem. J.* 1913, **49**, 87–116.

2. *Water-insoluble aldehydes and ketones*

The carbonyl compound (a known weight) is dissolved in ethyl alcohol (95%, free from aldehydes and ketones) and the solution made up (to 100 ml.). Aliquot portions (10 ml.) of the solution are added dropwise to the reagent and the estimation carried out as above. The reaction mixture should be allowed to stand for 2–24 hr. before the derivative is filtered off.

PREPARATION OF THE OXIME AND ESTIMATION OF EITHER THE HYDROCHLORIC ACID OR THE WATER FORMED

$$\frac{R}{R'}\!\!>\!\!C{=}O + H_2NOH.HCl \rightleftharpoons \frac{R}{R'}\!\!>\!\!C{=}NOH + H_2O + HCl.$$

Hydroxylamine hydrochloride is used widely as a quantitative reagent for the estimation of carbonyl compounds. The hydrochloric acid liberated, or the water formed, is determined by titrating the former with alkali and the latter with the Karl Fischer reagent (see pp. 50–5). The result is given sometimes as the hydroxyl number which is defined as the number of milligrams of hydroxylamine needed for the complete oximation of 1 g. of the carbonyl compound.[1]

The formation of oximes is a reversible reaction, and an equilibrium is established which is affected markedly by the acidity of the solution. It is therefore impossible theoretically to determine the percentage of the carbonyl compound present, but experimentally it is possible to arrange the conditions so that oximation is complete for all practical purposes. When low values result from an unfavourable equilibrium the oxime formation can usually be increased by adding either pyridine or calcium carbonate to the reaction mixture.[2] A procedure using the former in all determinations has been described as giving reliable results (see below).

[1] R. C. Stillman and R. M. Reed, *Perfumery Essent. Oil Record*, 1932, **23**, 278–86.
[2] W. M. D. Bryant and D. M. Smith, *J. Amer. Chem. Soc.* 1935, **57**, 57–61; E. R. Alexander and D. C. Dittmer, *ibid.* 1951, **73**, 1665–8; S. Sabetay, *Bull. Soc. Chim.* 1938, [5], **5**, 1419.

The methods based on the estimation of hydrochloric acid liberated in the reaction are subject to interference from other basic and acidic substances. This difficulty is overcome in a potentiometric method of estimation and is not associated with the Karl Fischer reagent for the determination of the water formed, although in the latter a correction must be made for any water or alkali present.

A. Methods for the Estimation of the Hydrochloric Acid

I. A large number of papers on the acidimetric method have been published proposing modifications of the original method to obtain accurate results with particular types of carbonyl compounds.[1] A method based on the work of Bennett and his co-workers is probably the best of the general methods which have been described (see below), but when it is applied to aldehydes and ketones of widely varying constitution the results are not always accurate. Slight modifications to the general procedure, however, overcome the difficulties and reliable results can usually be obtained. In the estimation of the percentage of an aldehyde in an essential oil the addition of a little benzene as solvent for the non-aldehydic portion of the oil is often necessary. With ketones the reaction is slow under these conditions, and the oxime is formed incompletely; but by using a solution of hydroxylamine hydrochloride in 80 % alcohol and heating the reaction mixture at 70° C. accurate values can be obtained.[2]

The hydrochloric acid formed is titrated with an alcoholic solution of sodium, or potassium, hydroxide using methyl orange, screened methyl orange (methyl orange-xylene cyanol)

[1] A. Brochet and R. Cambrier, *Compt. rend.* 1895, **120**, 449–54; L. Palfray and S. Tallard, *ibid.* 1934, **199**, 296–8; A. H. Bennett, *Analyst*, 1909, **34**, 14–17; A. H. Bennett and T. T. Cocking, *ibid.* 1931, **56**, 79–82; C. T. Bennett and M. S. Salamon, *ibid.* 1927, **52**, 693–5; '6th Report of the Essential Oil Sub-Committee', *ibid.* 1930, **55**, 109–10; M. E. Martin, K. L. Kelly and M. W. Green, *J. Am. Pharm. Assoc.* 1946, **35**, 220–3; H. Schultes, *Z. Angew. Chem.* 1934, **47**, 258–9; S. Sabetay, *Bull. Soc. Chim.* 1938, [5], **5**, 1419–22; J. G. Malty and G. R. Primavesi, *Analyst*, 1949, **74**, 498–502; A. M. Trozzolo and E. Lieber, *Anal. chem.* 1950, **22**, 764–8.
[2] A. H. Bennett and T. T. Cocking, *loc. cit.* (p. 145, n. 1); S. Sabetay, *loc. cit.* (p. 145, n. 1).

or bromophenol blue as indicator. The end-point is not sharp, and in making a choice of indicator experience in their use and ability to judge accurately their colour change from red to yellow or from blue to yellow are the deciding factors. The estimation has been done electrometrically using a glass electrode and the normal calomel cell, and the titration curves show that a sharp end-point is not possible with an indicator such as bromophenol blue.[1]

The use of a salt other than the hydrochloride has been tried successfully with a few carbonyl compounds.[2] In the estimation of citronellal the formation of hydrochloric acid affects the aldehyde content adversely and the result is low, but if the hydrochloride is replaced by the acetate the determination is reliable.

A mixture of hydroxylamine hydrochloride (30 ml.) and the carbonyl compound (sufficient to react with one-half and not more than two-thirds of the hydroxylamine hydrochloride) is left to stand in a stoppered bottle at room temperature, or when oximation is slow at 100° C., for 2 hr. When it is necessary to heat the reaction mixture the titration must not be carried out until the temperature returns to within 5° C. of room temperature (about 1 hr.). The hydrochloric acid is titrated with an alcoholic solution of sodium (or potassium) hydroxide (0·5 N), using methyl orange or bromophenol blue as indicator.

When an 80 % alcoholic solution is used methyl orange is unsuitable and must be replaced by dimethyl yellow (dimethyl-aminoazobenzene), which has a colour change the same as, and the pH range similar to, methyl orange.[3] If it is necessary the alcoholic mixture can be heated to 70° C. when oximation is generally complete within 60 min. The end-point is fairly distinct, and reproducible results can be obtained by the same person but concordant results are not always possible with different workers.

A blank titration should be carried out under the same

[1] A. Eitel, *J. prakt. Chem.* 1942, **159**, 292–303; cf. G. A. Pevtzov, *J. Appl. Chem. Russ.* 1943, **16**, 363; J. G. Malty and G. R. Primavesi, *loc. cit.* (p. 145, n. 1).
[2] H. I. Waterman and E. B. Elsbach, *Rec. trav. chim.* 1929, **48**, 1087–91.
[3] A. H. Bennett and T. T. Cocking, *loc. cit.* (p. 145, n. 1).

conditions, and the difference in the two titration values is a measure of the hydroxylamine which has combined with the carbonyl compound.

Accuracy $\pm 1\%$.

PREPARATIONS: *Indicator.* A 0.2% aqueous solution of methyl orange or a 0.4% alcoholic solution of bromophenol blue is used.

Alcoholic alkali. Sodium, or potassium, hydroxide is dissolved in aqueous methyl alcohol (60% of methyl alcohol, by volume unless oximation is difficult when 80% is used) to give a 0.5N solution. The standardization is made against hydrochloric acid (0.5N) using methyl orange as the indicator and running alkali into acid until the full yellow colour is obtained.

Hydroxylamine hydrochloride. Commercial hydroxylamine hydrochloride frequently contains impurities which make the end-point indeterminate. The pure hydrochloride (3.475 g. recrystallized from water) is dissolved in aqueous methyl alcohol (95 ml., 60% by volume), methyl orange is added (10 drops) and the full yellow colour of the indicator adjusted by adding alcoholic alkali (0.5N). The solution is then made up to the mark (100 ml.) with aqueous methyl alcohol (60% by volume).

II. An alternative to the above method has been tried with a large number of aldehydes and ketones of widely varying constitution and a satisfactory general procedure has been described.[1] The essential difference is in the addition of pyridine to the reaction mixture which forces the equilibrium of the oxime formation nearer to completion and also makes the reaction mixture initially neutral to the indicator. A blank determination is, therefore, unnecessary, but it is better to do at least one such titration under the conditions by which the reaction mixture is estimated and to determine the end-points by comparison. The pyridine has a slight buffer action which makes the indicator colour change less sensitive, but its value in completing oximation when the equilibrium is unfavourable

[1] W. M. D. Bryant and D. M. Smith, *loc. cit.* (p. 144, n. 2).

more than compensates for the diminished sensitivity. The reaction is rather slow and cannot be used for very low concentrations of aldehydes and ketones.

The method is applicable to aliphatic, aromatic and alicyclic aldehydes and ketones but not to compounds such as amides. Structural peculiarities have a marked effect on the rate of reaction, but they are seldom of sufficient importance to make the method inapplicable.

A solution of hydroxylamine hydrochloride (30 ml., see below) is added to a mixture of pyridine and bromophenol blue (100 ml.) in a pressure bottle. The colour should be greenish blue after mixing. The carbonyl compound (sufficient to react with one-half and not more than two-thirds of the hydroxylamine hydrochloride), is added and the mixture left to stand either at room temperature for $\frac{1}{2}$–48 hr., or in a water bath at 100° C. for 2 hr., after which time equilibrium is usually established. When heating is necessary the mixture must be allowed to cool to within 5° C. of the temperature of the room before the titration is done (about 1 hr.). The colour of the indicator changes with heating, becoming yellow at high temperatures; it is important, therefore, to carry out the titration of the reaction mixture and the blank determination at the same temperature and to avoid unnecessarily long heating which causes the indicator to deteriorate. Satisfactory results can be expected after heating at 100° C. for 5 hr. The free hydrochloric acid is determined by titration with alcoholic sodium, or potassium, hydroxide (0·5 N) until the colour at the end-point matches the blank estimation.

It is important to keep the volumetric ratio of the reagent to the pyridine solution constant in order to maintain the sensitive green colour of the indicator in the blank estimation. The colour is largely dependent upon the amount of water present in the final mixture, and the introduction of appreciable quantities of water over those included in the solutions of the reagents increases the alkaline buffer properties of the pyridine, and so interferes with the acidimetric titration. Up to 5 ml. of water can be added to the reaction mixture if a similar amount is added to the blank estimation.

Percentage of $>$CO compound

$$= \frac{\text{ml. NaOH} \times \text{N} \times \text{mol. wt.} \times 100}{\text{wt. of carbonyl compound} \times 1000}.$$

Accuracy $\pm 1\%$.

PREPARATIONS: *Indicator*. The indicator is included in the pyridine solution which is made up by adding an alcoholic solution of bromophenol blue (0·25 ml., 4%) to pyridine (20 ml.) and diluting (to 1 l.) with ethyl alcohol (95%).

Alcoholic alkali. Sodium or potassium hydroxide is dissolved in aqueous methyl alcohol (90%) to give a 0·5N solution. The standardization is as above.

Hydroxylamine hydrochloride. The pure hydrochloride (35 g., see above) is dissolved in distilled water (160 ml.) and the solution made up (to 1 l.) with ethyl alcohol (95%). The reagent should not be strongly acid. A solution (10 g.) in distilled water (50 ml.) should be neutralized by less than 8 ml. of alcoholic sodium hydroxide (0·5N).

III. In the potentiometric method the carbonyl compound can be estimated in the presence of an acid.[1] The pH of the hydroxylamine solution is adjusted to an apparent value of 2·50 at which point the titration of hydrochloric acid is stoichiometrical and carboxylic acids scarcely interfere. Mineral acids and organic acids of comparable strength (e.g. sulphonic acids) titrate quantitatively under these conditions and are estimated in a preliminary potentiometric titration. Hydroxylamine hydrochloride is then added, and after standing for 15 min. the free hydrochloric acid is determined.

The carbonyl compound (containing up to 25 m.equiv.) is weighed into a beaker (400 ml.) containing the reagent (100 ml., see below). After standing at room temperature for 15 min. the solution is titrated potentiometrically to pH 2·50 with a methyl alcoholic solution of sodium hydroxide (0·5N, see below). The mixture is diluted (to 150 ml.) with methyl alcohol (90%) and the apparent pH changes to about 2·65. The titration is continued drop by drop until the pH is the same as that of a blank

[1] D. M. Smith and J. Mitchell, Jr., *Anal. Chem.* 1950, **22**, 750–5.

made up from the reagent (100 ml.) and ethyl alcohol (50 ml., 90 %) which has an apparent pH of approximately 2·65.

PREPARATIONS: *Hydroxylamine hydrochloride.* Pure hydroxylamine hydrochloride (35·0 g.) is dissolved in distilled water (100 ml.) and diluted (to 1 l.) with ethyl alcohol (95 %). A solution of thymol blue is added (6 ml., 0·1 %, see below) and the pH is adjusted (to 2·50 ± 0·01) with aqueous hydrochloric acid (6 ml., 0·5 N).

Thymol blue. Thymol blue (1 g.) is powdered and made into a paste by grinding with sodium methoxide in methyl alcohol (4 ml., 0·5 N). The solution is made up (to 1 l.) with dry methyl alcohol.

Sodium hydroxide in methyl alcohol. Sodium hydroxide (80 g., pellets) is dissolved in methyl alcohol (to make 4 l. containing 10 % of water). A solution of sodium hydroxide in methyl alcohol has a significant coefficient of expansion, and the temperature at the time of titration should be noted and, if necessary, a correction should be made to the normality. The reagent is standardized against hydrochloric acid (0·5 N) or against benzoic acid.

B. Estimation of the Water

A successful estimation of the carbonyl content depends upon the complete reaction of the aldehyde or ketone with a reagent followed by an accurate estimation of one of the products of the reaction. The first condition is satisfied for all practical purposes by preparing the oxime using a large excess of hydroxylamine hydrochloride, but this prevents the direct titration of the water formed by the Karl Fischer reagent. The oximes themselves do not react with the Karl Fischer reagent, but the hydroxylammonium ions do and they must be stabilized after oximation is finished and before the final titration.[1] An excess of sulphur dioxide and pyridine in methyl alcohol is added to the reaction mixture 10 min. before the titration to convert the excess of hydroxylamine hydrochloride to the pyridine salt of sulphamic acid:[2]

$$H_2NOH.HCl + SO_2 + 2C_5H_5N = H_2N.SO_3NH.C_5H_5 + C_5H_5N.HCl.$$

[1] J. Mitchell, Jr., D. M. Smith and W. M. D. Bryant, *J. Amer. Chem. Soc.* 1941, 63, 573–4; cf. *ibid.* 1940, 62, 3504–5.
[2] Cf. F. Raschig, *Ber.* 1887, 20, 584–9.

The titration is easy and rapid once the difficulty of using the Karl Fischer reagent has been overcome (see pp. 50–5). The determination is based on the water formed in the preparation of the oxime, so that a titration of the carbonyl compound and of the hydroxylamine hydrochloride solution must be done to find the water they contain. The method is used for all types of carbonyl compounds of the aliphatic, aromatic and alicyclic series although camphor is an exception as it fails to react. Acids, ethers and hydrocarbons do not interfere with the estimation, but organic acids react in the presence of an esterifying catalyst to form water which is estimated in the titration. This error can be eliminated by substituting pyridine for methyl alcohol in the preparation of the sample. Free alkali titrates, molecule for molecule, as water, and it is essential therefore to make a correction for any alkali present. A large concentration of an amine, which does not react with the carbonyl compound, can be prevented from interfering by adding an excess of acetic acid.

The carbonyl compound (5–10 ml. of a solution in dry methyl alcohol containing not more than 0·1 mol. of the carbonyl group or water) is weighed into a volumetric flask (100 ml.), and a solution is made up by diluting to the mark with dry methyl alcohol (from a supply kept in a thermostat at 25° C.). A solution of hydroxylamine hydrochloride (30 ml., 0·5N, see below) and pyridine (5 ml.) are run into a glass-stoppered flask (250 ml.) from a rapid-delivery burette or pipette and a measured volume (10 ml.) of the solution of the carbonyl compound is added. The flask is placed in a water bath at $60 \pm 1°$ C., the stopper is loosened momentarily to allow for the expansion of air and then tightened. The flask is heated for 2 hr. and then cooled to room temperature. A solution of sulphur dioxide (25 ml., 1M, see below) in pyridine and methyl alcohol is added, the mixture is left for 10–60 min. and then titrated with the Karl Fischer reagent (see pp. 50–5).

One blank determination at least must be made on the dry methyl alcohol and on the solution of the carbonyl compound. A dry pyridine solution of hydrogen cyanide (2 %) is added to the latter before titration, as the carbonyl group must be protected.

Accuracy $\pm 1 \%$.

PREPARATIONS: *The Karl Fischer reagent* (see p. 55).

Hydroxylamine hydrochloride. The pure hydrochloride (35 g.) is dissolved in methyl alcohol (less than 0·1% water) and diluted (to 1 l.).

Sulphur dioxide. Liquid sulphur dioxide (45 ml.) is added to a chilled solution of pure, dry pyridine (80 ml., see p. 62) in dry methyl alcohol (875 ml.).

MISCELLANEOUS

(1) Aldehydes, similar to those found in essential oils, have been estimated by heating under a reflux condenser for $2\frac{1}{2}$ hr. with potassium benzylate. One mol. of acid is formed from each mol. of the aldehyde, and the amount is determined by titrating the excess alkali with standard acid using phenolphthalein as indicator and subtracting the value from that obtained in a control experiment. A Pyrex apparatus is recommended. The method is not applicable to aliphatic aldehydes.[1]

(2) The percentage of carbonyl group in a limited number of aldehydes, including cinnamic, anisic, vanillin, heliotropin, and benzaldehyde, has been estimated by titrating with benzidine acetate using blood and hydrogen peroxide as the indicator. The standard solution is prepared by dissolving benzidine (9·2 g.) in glacial acetic acid (180 ml.) and diluting (to 1 l., 0·1 N solution).

Blood (1 ml.) in glacial acetic acid (5 ml.) is added to the aldehyde (0·2–0·3 g.) in the same solvent (5 ml.) and the mixture titrated with the standard solution of benzidine acetate. The end-point is reached when a drop of the solution gives a blue colour on a strip of paper moistened with hydrogen peroxide.[2]

(3) Indirect methods of estimation of the percentage of the carbonyl group in carbonyl compounds have been described involving the quantitative preparation of a derivative followed by its estimation:

(a) The nitrogen content of semicarbazones (0·1–0·2 g.) can be found by heating under a reflux condenser for 30 min. with water (100 ml.) and concentrated sulphuric acid (10 ml.), adding

[1] L. Palfray, S. Sabetay and D. Sontag, *Compt. rend.* 1932, **194**, 1502–5.
[2] P. N. van Eck, *Pharm. Weekblad.* 1928, **65**, 82–4.

potassium iodate (0·4–0·5 g.) and boiling until the colour of
iodine disappears. The mixture is diluted with water (200 ml.)
made alkaline with sodium hydroxide solution (50 ml., 10 N) and
distilled (until 150 ml. are collected). The distillate is titrated as
in the Kjeldahl estimation of nitrogen (see pp. 190–5). Only
one-third of the total nitrogen present in the semicarbazone is
estimated in this method.[1]

(b) An alternative to the above method hydrolyses the semicarbazone with a mixture of dilute hydrochloric acid (15 %) and
mercuric chloride (5 %).[2] The semicarbazide formed on hydrolysis
is decomposed into ammonia and hydrazine, and the latter is
oxidized by the mercuric chloride.[3] The ammonia, representing
one-third of the total amount of nitrogen in the semicarbazone,
can be determined as described under nitro compounds (see
pp. 190–5) or by Parnas and Wagner's modified micro-Kjeldahl
method.[4] The mercury ammonium complex is decomposed and
the reaction mixture made alkaline with a saturated solution of
sodium thiosulphate diluted with an equal volume of sodium
hydroxide (40 %).

It is unnecessary to isolate the pure semicarbazone, but the
excess of semicarbazide and any impurities present which yield
ammonia on acid hydrolysis must be removed. The separation
of the semicarbazone from the unreacted reagent may be carried
out by

(i) precipitation of the semicarbazone with water,

(ii) evaporating the mixture to dryness and removing the
semicarbazide by washing with water,

(iii) evaporating the mixture to dryness and extracting the
residue with ether, the ethereal solution being shaken with
water to remove the semicarbazide.

(c) p-Carboxyphenylhydrazine has been described as a useful
reagent for the detection of carbonyl groups in unknown substances; its derivatives can be used for the estimation of the

[1] S. Veibel, Bull. Soc. chim. France, 1927, [4], 41, 1410–16.

[2] R. P. Hobson, J. Chem. Soc. 1929, pp. 1384–5; cf. S. Veibel, ibid. 1929,
p. 2423.

[3] C. Maselli, Gazzetta, 1905, 35, [i], 267.

[4] J. Grant, Quantitative Organic Micro-analysis based on the methods of
Fritz Pregl, 5th English ed., 1951, pp. 95–104.

percentage of carbonyl group present. The substituted hydrazone is dissolved in ethyl alcohol, and the free carboxyl group is titrated with baryta using phenolphthalein as indicator. An accuracy of about 0·5 % is claimed.[1]

(d) Certain halogenated and hydroxy aldehydes react as monobasic acids and can be titrated with alkali hydroxides in aqueous or alcoholic solution using phenolphthalein, methyl orange or Poirrier's blue as indicator.[2] Chloral hydrate, chloral alcoholate and bromal are neutralized by one equivalent of alkali. Hydroxybutyraldehyde and aldehydic sugars are neutral towards all three indicators, but whereas salicylaldehyde, p-hydroxybenzaldehyde and vanillin are neutral towards methyl orange, they are monobasic towards phenolphthalein and Poirrier's blue.

(4) Citral, and aldehydes of the terpene class, may be estimated through the thiosemicarbazone which can be prepared quantitatively.[3]

Thiosemicarbazide (0·8 g.) is added to the aldehyde (2·3 g.) in alcohol (50 ml.), and the mixture heated on a steam bath until the reagent is dissolved. The solution is evaporated and the residue dried by drawing a stream of air through the flask immersed in a boiling water-bath. The thiosemicarbazone is extracted by adding carbon disulphide (150 ml.) and heating in a water-bath for a few minutes. The solution is filtered, the precipitate washed with carbon disulphide, and the filtrate allowed to evaporate. The residue is dried at 50° C., left to cool in a desiccator, and weighed.

(5) An excess of 2 : 4-dinitrophenylhydrazine in hydrochloric acid (2N) is added to the carbonyl compound and when reaction is complete titanous chloride (1·5 mols. per mol. of dinitrophenylhydrazine) is added and the excess of the reagent

---

[1] S. Veibel and N. Hauge, *Bull. Soc. chim. France*, 1938, [5], **5**, 1506–9; S. Veibel, *Acta chem. Scand.* 1947, **1**, 54–68; S. Veibel and H. W. Schmidt, *ibid.* 1948, **2**, 545–9.

[2] A. Astruc and H. Murco, *Compt. rend.* 1900, **131**, 943–5; H. Meyer, *Monatsh.* 1903, **24**, 832–9.

[3] L. G. Radcliffe and W. J. N. Swann, *Perfumery Essent. Oil Record*, 1928, **19**, 47–51.

is determined by titration with ferric alum using potassium thiocyanate as the indicator.[1]

(6) The addition of the carbonyl compound to an excess of hydroxylamine and titration of this excess with hydrochloric acid after oximation has been completed has been recommended for the estimation of aldehydes and ketones. The hydroxylamine is prepared from the hydrochloride and sodium acetate, and bromophenol blue is the indicator used.[2]

(7) An excess of hydrazine sulphate is added to an aromatic aldehyde, the solution is filtered, made up to a definite volume and left to stand for several hours. An excess of iodine is added to a measured volume, the mixture is made alkaline with sodium hydroxide and the excess iodine is titrated with a standard solution of sodium thiosulphate.[3]

(8) An alternative method to the above (see para. (6) above) uses an excess of hydroxylamine for the formation of the oxime followed by an estimation of the excess. Accurate values can be obtained if precautions are taken to avoid loss of hydroxylamine.[4]

(9) Aldehydes can be oxidized to the corresponding acid with silver oxide and so may be estimated by titrating the reaction mixture with a standard solution of sodium hydroxide using phenolphthalein as the indicator.[5]

(10) Acetaldehyde has been estimated by oxidation with an excess of precipitated silver oxide followed by a determination of the excess. The reaction can be accelerated by adding an alkali hydroxide and heating the reaction mixture when the concentration of aldehyde has become small. The unused silver oxide is estimated by titration with potassium iodide.

[1] W. Schöniger and H. Lieb, *Mikrochemie ver. Mikrochim. Acta*, 1951, **38**, 165–7.
[2] H. I. Waterman and E. B. Elsbach, *Rec. trav. chim.* 1929, **48**, 1087–91; R. C. Stillman and R. M. Reed, *Perfumery and Essent. Oil Record*, 1932, **23**, 278–86.
[3] L. Lautenschläger, *Arch. Pharm.* 1918, **256**, 81; cf. L. Fuchs and O. Matzke, *Scientia Pharm.* 1948, **17**, 1–11.
[4] A. H. Bennett, *Analyst*, 1909, **34**, 14; A. H. Bennett and F. K. Donovan, *ibid.* 1922, **47**, 146; H. I. Waterman and E. B. Elsbach, *loc. cit.*
[5] J. Mitchell, Jr., and D. M. Smith, *Anal. Chem.* 1950, **22**, 746–50.

Uniformity of operation is difficult and the method is not applicable to slowly reducing aldehydes (e.g. benzaldehyde) or to small quantities of acetaldehyde.[1]

(11) 5:5-Dimethyl*cyclo*hexa-1:3-dione (also known as dimethyldihydroresorcinol, dimedone, dimetol, methone and metol) is well known as a useful reagent for the characterization of aldehydes.[2] The derivative is formed quantitatively and has been used to a limited degree for the gravimetric estimation of aldehydes.[3] The reaction is specific for aldehydes as no condensation takes place with ketones.

The aldehyde (25 ml. of a solution containing about 0·1 g. per l.) is added to an aqueous solution of dimedone (10 % excess of a saturated solution which contains about 0·004 g. per ml.) and a buffer solution (250 ml., 2 vol. of N sodium acetate, 1 vol. N hydrochloric acid) and the mixture is left to stand, with occasional shaking, at room temperature for 24 hr. The solid which is formed is filtered into a Gooch crucible, washed with water and dried to constant weight at 55–60° C.

(12) A colorimetric method of estimation for the carbonyl group is based on the intense red colour produced when a solution of sodium, or potassium, hydroxide is added to an alcoholic solution of a 2:4-dinitrophenylhydrazone. The reaction is very sensitive and has been used for the determination of keto-steroids in biological extracts in concentrations as low as 10^{-4} to 10^{-6} M.[4]

(13) A number of the higher aldehydes have been estimated colorimetrically with a modified Schiff's reagent after extraction with petrol ether. The sensitivity is claimed as twenty times that of ordinary Schiff's, but no figures are given.[5] The method has been used by other investigators and found to be useful.

[1] W. Ponndorff, *Ber.* 1931, **64**, 1913–23.
[2] See F. Wild, *Characterisation of Organic Compounds*, pp. 135–7.
[3] J. H. Yoe and L. C. Reid, *Ind. Eng. Chem.* (Anal. ed.), 1941, **13**, 238–40.
[4] L. C. Clarke, *Anal. Chem.* 1951, **23**, 541–2.
[5] H. Schibsted, *Ind. Eng. Chem.* (Anal. ed.), 1932, **4**, 204–8.

SPECIFIC FOR PARTICULAR ALDEHYDES
FORMALDEHYDE
Methods of estimation

I. The oxidation of formaldehyde with hydrogen peroxide in the presence of a standard solution of sodium hydroxide is the method described in the *British Pharmacopoeia*:[1]

$$H.CHO + H_2O_2 = H.COOH + H_2O,$$
$$H.COOH + NaOH = H.COONa + H_2O.$$

The formic acid is neutralized as it is formed, and the excess alkali titrated with sulphuric acid (N) using phenolphthalein as the indicator. After oxidation is completed, which takes at least 90 min. at room temperature, and before titration, the reaction mixture must be warmed to remove hydrogen peroxide in order to prevent the oxidation of the phenolphthalein to a substance which remains colourless in alkaline solution.

II. The oxidation of formaldehyde and determination of the excess of the oxidizing agent has been used as a means of estimation. The method is most suitable for small amounts of pure formaldehyde, but if large quantities are used an error is possible from the formation of an aldehyde-iodine compound.

(a) A mixture of iodine and sodium hydroxide solutions is added to a solution of formaldehyde. After leaving to stand for 10 min. the mixture is acidified with hydrochloric acid and the excess free iodine is titrated with sodium thiosulphate.[2] Accurate results are obtained and methyl alcohol and acetic acid do not interfere.

(b) An alkaline solution of formaldehyde may be oxidized with bromine and iodine equivalent to the excess of bromine can be liberated from the acidified solution and titrated with sodium

[1] O. Blank and H. Finkenbeiner, *Ber.* 1898, **31**, 2979–81; A. Foschini and M. Talenti, *Z. anal. Chem.* 1939, **117**, 94–9; W. Meyer, *Pharm. Ztg*, 1929, **74**, 771–3; M. Maccormac and D. T. A. Townend, *J. Chem. Soc.* 1940, pp. 151–6; J. Büchi, *Pharm. Acta Helv.* 1931, **6**, 1.
[2] G. Romijn, *Z. anal. Chem.* 1897, **36**, 18–24; R. Signer, *Helv. Chim. Acta*, 1930, **13**, 43–6; J. Büchi, *loc. cit.* (p. 157, n. 1); R. Wolf, *Rev. brasil. quím.* (São Paulo), 1943, **15**, 159–62.

thiosulphate.[1] A standard solution of potassium bromide-potassium bromate is used for the production of the bromine.

(c) The method described under benzaldehyde (see p. 165) which uses chloramine-T as an oxidizing agent can be used for formaldehyde.

III. Formaldehyde can be estimated by adding an excess of a neutral, aqueous solution of sodium sulphite and titrating the sodium hydroxide which is formed (cf. p. 134):[2]

$$H.CHO + Na_2SO_3 + H_2O = H.CH(OH).SO_3Na + NaOH.$$

An aqueous solution of formaldehyde (25 ml., containing about 1·5 g. of formaldehyde) is neutralized with sodium hydroxide (N) using thymolphthalein as the indicator (3 drops, 1%). A neutral solution (to thymolphthalein) of sodium sulphite (6·5 g. crystalline) in water (25 ml.) is added to the formaldehyde solution, and the mixture is titrated slowly with sulphuric or hydrochloric acid (2N). Accurate results are obtained in the presence of high concentrations of ethyl alcohol, but acetic acid in appreciable quantity gives high values.

IV. Several methods are available for the colorimetric estimation of formaldehyde, including:

(a) Schryver's method. When a dilute solution of formaldehyde phenylhydrazone is treated with potassium ferricyanide in the presence of an excess of hydrochloric acid an intense magenta colour results, the depth of which is directly proportional to the concentration of the aldehyde phenylhydrazone. The method has been used for estimating formaldehyde in air as it can be used for concentrations between 1 and 5 parts of formaldehyde per million, but outside these concentrations the colour is either too faint or too intense for an accurate comparison with a standard solution.[3]

[1] L. Spitzer, *Chem. Ztg.* 1933, **57**, 224.

[2] G. Lemme, *Chem. Ztg.* 1903, **27**, 896; K. Täufel and C. Wagner, *Z. anal. Chem.* 1926, **68**, 25–33; J. Büchi, *loc. cit.* (p. 157, n. 1); L. Malaprade, *Compt. rend.* 1934, **198**, 1037–9 (cf. pp. 134–6).

[3] S. B. Schryver, 'On application of formaldehyde to meat', *Food Rep.* no. 9, 1909, Reps. of local government board on public health and medical subjects, new series, no. 12; R. W. Kersey, J. R. Maddocks and T. E. Johnson, *Analyst*, 1940, **65**, 203–6; E. H. Callow, *ibid.* 1927, **52**, 391–5; D. Matukawa, *J. Biochem.* (Japan), 1939, **30**, 385.

A measured volume of air is drawn through a solution of phenylhydrazine hydrochloride (10 ml., see below) in either a cylindrical wash-bottle fitted with a Jena glass distribution tube (0·75 in. dia.) with a fused sintered glass disk (type 33 c. G. 1, pore diameter 100–120μ) or in a similar apparatus. The air should be aspirated at about 10 l. an hour, and not exceed 40 l. in all, and should give a solution containing 1–5 mg. of formaldehyde per litre. The solution is diluted (to 50 ml.) with water, and a portion (10 ml.) is transferred to a Nessler tube (20 ml.). A freshly prepared aqueous solution of potassium ferricyanide (1·0 ml., 5 %) and hydrochloric acid (4 ml. of concentrated acid, $d = 1·16$, diluted to 20 ml. with water) are added, the solutions are mixed well and left to stand for 10 min. The colour is then compared in similar Nessler tubes with a series of standards made from the same reagents. Formaldehyde (1–5 p.p.m. see below) is added to a solution of phenylhydrazine hydrochloride (2 ml.) and water (7 ml.) and the tubes completed as above.

PREPARATIONS: *Phenylhydrazine* (1 g.) is suspended in water (about 5 ml.) and hydrochloric acid (2 ml., $d = 1·16$) is added. The solution is diluted, filtered and made up (to 100 ml.).

Standard solution of formaldehyde. Formaldehyde (25 ml., commercial 40 % weight/volume) is diluted (to 1 l.). If great accuracy is necessary the solution can be standardized by Ripper's method, otherwise 1 ml. of the solution can be taken as equivalent to 0·01 g. of formaldehyde.

(*b*) A method of estimating formaldehyde in dilute aqueous solution is based on the colour reaction with a phloroglucinol reagent; the colour is compared after 3 min. with the standard mixtures of congo red and methyl orange.[1] An alternative colour standard using mixtures of alizarin yellow *R* and methyl orange and making the comparison after 5 min. has been found to give better results.[2]

[1] R. J. Collins and P. J. Hanzlik, *J. Biol. Chem.* 1916, **25**, 231–7; P. J. Hanzlik, *ibid.* 1920, **42**, 411–13; cf. R. C. Hoather and P. G. T. Hand, *Analyst*, 1939, **64**, 29–30.
[2] R. P. Spoto, *Diagnostica Tec. Lab. (Napoli) Riv. mensile*, 1931, **2**, 362.

(c) The violet-red colour produced by dissolving difurfurylideneacetone in sulphuric acid (60 %) fades on the addition of aliphatic aldehydes. The rate at which the colour is discharged is proportional to the concentration of the aldehyde.[1]

(d) Formaldehyde in concentrations between $2 \cdot 5 \times 10^{-3}$ and 30×10^{-3} can be estimated by mixing a saturated aqueous solution of β-naphthol (5 ml.) with the aldehyde solution and then adding concentrated sulphuric acid (5 ml., $d = 1 \cdot 84$), so that a zone of separation appears. The reddish tint is compared with standards.[2]

V. Potassium cyanide reacts quantitatively with formaldehyde and any excess can be estimated by titration with silver nitrate or by reaction with bromine water.[3]

A solution of formaldehyde (10 ml., containing between 10 and 50 mg. of formaldehyde) is added to a mixture of potassium cyanide (50 ml., $0 \cdot 1$N) and magnesium sulphate (about 5 ml., 30 %) which acts as a catalyst. The mixture is stirred vigorously, and after 1 or 2 min. ammonium chloride (1 g.) is added to dissolve the magnesium hydroxide. Ammonium hydroxide (2–3 drops, 6N) and a little potassium iodide are added as the indicator, and the mixture is then titrated rapidly with silver nitrate ($0 \cdot 1$N) until a turbidity of silver iodide persists for 1–2 min.

If less than 10 mg. of formaldehyde is present better results are obtained as follows:

A solution of formaldehyde (10 ml.) is added to potassium cyanide (10–25 ml., $0 \cdot 1$N) with stirring. After allowing to stand for 3 min. the solution is made strongly acid with concentrated hydrochloric acid, and bromine water is added until the yellow colour is permanent. Phenol (1–2 g.) is added to react with the excess bromine, and after adding potassium iodide (10 ml., 10 %) the mixture is titrated with sodium thiosulphate.

[1] V. V. Chelintsev and E. K. Nikitin, *J. Gen. Chem.* (U.S.S.R.), 1937, **7**, 2324–31.
[2] J. M. Hambersin, *Bull. soc. chim. Belg.* 1937, **46**, 519–24.
[3] E. Schulek, *Ber.* 1925, **58**, 732–6.

VI. The reaction between formaldehyde and ammonia to give hexamethylene tetramine has been used for the estimation of the aldehyde.[1] Two methods have been described:

(a) A solution of formaldehyde is heated with an excess of aqueous ammonia in a pressure bottle, and after allowing the bottle to cool the excess ammonia is titrated with a standard solution of an acid.[2]

(b) A mixture of formaldehyde, ammonium chloride and sodium hydroxide is allowed to stand until the formation of hexamethylene tetramine is completed. The excess alkali is then titrated with acid (0·5 N), using cresol red or thymol blue as the indicator. Four mol. of sodium hydroxide are equivalent to 1 mol. of formaldehyde.

VII. Dimethyldihydroresorcinol (also known as dimedone, 5:5-dimethyl*cyclo*hexa-1:3-dione, dimetol, methone, and metol) reacts quantitatively with formaldehyde to give a colourless, crystalline precipitate which can be estimated either gravimetrically (cf. p. 156),[3] or by titration with sodium hydroxide.[4] The gravimetric method has been used for microdeterminations.[5]

VIII. A method described as 'sensitive and specific' is based on the precipitate formed from the reaction between formaldehyde and trypaflavine in the presence of hydrochloric acid.[6]

IX. A nephelometric method of estimation for formaldehyde has been developed using the methyleneaniline formed by reaction with aniline.[7]

[1] L. Leglers, *Ber.* 1883, **16**, 1333–7; A. G. Craig, *J. Amer. Chem. Soc.* 1901, **23**, 638–43; Z. Peška, *Chem. Ztg*, 1901, **25**, 743; J. H. Norris and G. Ampt, *Proc. Soc. Chem. Ind. Victoria*, 1933, **33**, 801–10; A. Foschini and M. Talenti, *Z. anal. Chem.* 1939, **117**, 94–9.

[2] L. Leglers, *loc. cit.* (p. 161, n. 1); A. G. Craig, *loc. cit.* (p. 161, n. 1); Z. Peška, *loc. cit.* (p. 161, n. 1).

[3] M. V. Ionescu and C. Bodea, *Bull. Soc. chim. France*, 1931, [10], **47**, 1408–19; P. Fleury and J. Lange, *J. pharm. chim.* 1933, **17**, 196–208; J. H. Yoe and L. C. Reid, *Ind. Eng. Chem.* (Anal. ed.), 1941, **13**, 238–40.

[4] D. Vorländer, C. Ihle and H. Volkholz, *Z. anal. Chem.* 1929, **77**, 321–7.

[5] T. Uchino and I. Hosaka, *Japan J. Exp. Med.* 1938, **16**, 227–37.

[6] S. Ohyama, *ibid.* 1935, **13**, 327–30.

[7] G. Toussaint, J. Detrie and M. Verain, *Compt. rend. soc. biol.* 1934, **117**, 193–4.

X. The reduction of silver nitrate to silver by formaldehyde has been used as a method of estimation. The precipitated silver is filtered off, washed with water and dissolved in warm, dilute nitric acid (1 to 3 parts water) and estimated by titration with ammonium thiocyanate (0·1N), using ferric alum as the indicator.[1]

XI. Nessler's solution has been used for the estimation of formaldehyde alone or in the presence of ketones and keto aldehydes.[2]

Nessler's solution (10 ml., 2·8 % mercuric chloride and 7·5 % potassium iodide) and an aqueous solution of sodium hydroxide (6 ml., 2N) are added to a solution of formaldehyde (10 ml., containing less than 0·15 g. of formaldehyde). The mixture is stirred for 5 min. and then acetic acid (7·5 ml., 2N) and methyl alcohol are added. The precipitate is filtered, washed successively with methyl alcohol and water and then added to an excess of iodine (20 ml., 0·1N). The iodine remaining after the reaction is finished is titrated with a standard solution of sodium thiosulphate.

XII. A method which is very useful for the estimation of small concentrations of formaldehyde (sensitivity $0·3 \times 10^{-7}$M) is based on the purple colour produced when a solution of formaldehyde is heated with chromotropic acid (1:8-dihydroxynaphthalene-3:6-disulphonic acid).[3] A reaction takes place with other aldehydes, but the purple colour is specific for formaldehyde. Acraldehyde, glyceraldehyde, β-hydroxypropionaldehyde and acetaldehyde interfere, and compounds such as formic acid, acetic acid, oxalic acid, acetone, glycerol, glycose and mannose react with the reagent.

Formaldehyde (0·4–0·9 ml., containing not more than 100μg.) is added to an aqueous solution of chromotropic acid in a test-

[1] O. Heim, *Ind. Eng. Chem.* (Anal. ed.), 1929, **1**, 128; 1938, **8**, 431; E. E. Rebagliati, *Ann. Farm. Bioquim.* 1930, **1**, 150–8; Y. A. Fialkov and S. D. Shargorodskiĭ, *Mem. Inst. Chem. All-Ukrainian Acad. Sci.* 1934, **1**, 209–21.

[2] W. Stüve, *Arch. Pharm.* 1906, **244**, 540; J. Bolle, H. Jean and T. Jullig, *Mem. services chim. état* (Paris), 1948, **34**, 317–20; E. R. Alexander and E. J. Underhill, *J. Amer. Chem. Soc.* 1949, **71**, 4014–19.

[3] C. E. Bricker and H. R. Johnson, *Ind. Eng. Chem.* (Anal. ed.), 1945, **17**, 400–2.

tube (0·5 ml., 2·5 g. of the dry acid dissolved in 25 ml. of water and filtered). Concentrated sulphuric acid (5 ml., in small amounts) is poured down the side of the tube with continuous shaking, the mixture is heated in a boiling water bath for 1 hr. and then left to cool in cold water for 15 min. The intensity of the purple colour is measured and compared with standards in a Spekker photoelectric absorptiometer using Calorex 503 and yellow 606 Ilford filters.

XIII. Formaldehyde in aqueous solution may be estimated in a polarograph.[1]

ACETALDEHYDE

Methods of estimation

I. Acetaldehyde can be estimated accurately, even in dilute solution, by one of the following methods:

(*a*) Ripper's bisulphite method (see p. 131; cf. p. 134).[2]

(*b*) Oximation with hydroxylamine hydrochloride (see p. 144).[3]

II. Aldehydes can be estimated colorimetrically by comparing the colour restored to a decolorized solution of magenta with a standard. The alcohol used for making up the solutions must be freed from aldehydes before the method can be applied successfully (see p. 188).

A solution (50 ml.) is prepared by adding aldehyde-free aqueous alcohol (1:1 by volume) to an aqueous solution of the aldehyde (5–10 ml.). A decolorized solution of magenta (25 ml., see below) is added and the mixture allowed to stand for 15 min. The solution of unknown concentration is matched with standard solutions prepared under the same conditions using Nessler tubes, a colorimeter or a photoelectric absorptiometer. The solutions and the reagents should be kept at 15° C.

---

[1] M. J. Boyd and K. Bambach, *Ind. Eng. Chem.* (Anal. ed.), 1943, **14**, 314–15; E. C. Barnes and H. W. Speicher, *J. Ind. Hyg. Toxicol.* 1942, **24**, 10–17.

[2] Cf. J. Wagner, *Biochem. Z.* 1928, **194**, 441; S. Hähnel, *Svensk Kem. Tids.* 1935, **47**, 275.

[3] Cf. V. G. Shaposhnikov, Y. I. Makovskaya and N. A. Kalinicheva, *Trudui Gosudarst. Opuit. Zavoda Sintet Kauchuka Litera* B. III, *Synthetic Rubber*, 1934, 118–28; I. E. Filinov and V. P. Shatalov, *Sintet Kauchuk*, 1934, no. 1. 22–30.

PREPARATIONS: *Aldehyde-free alcohol.* See p. 188.

Decolorized solution of magenta. An aqueous solution of sulphur dioxide (0·5 g. an aqueous solution saturated at 20° C. contains 11·3 g. of sulphur dioxide in 100 ml.) is added to magenta (0·05 g.) dissolved in water (50 ml.); the mixture is diluted (to 100 ml.) and allowed to stand until colourless. The concentration of the sulphur dioxide solution can be determined by titration with a standard solution of iodine.

Standard solution of acetaldehyde. Acetaldehyde ammonia is purified by grinding several times in a mortar with dry ether and decanting the ether. The pure compound is dried in a current of air and then *in vacuo* over sulphuric acid. A solution of acetaldehyde is prepared by dissolving the solid (1·386 g.) in alcohol free from aldehydes (50 ml., 95%), adding alcoholic sulphuric acid (22·7 ml., N) and making the solution up to 100 ml. (allowing an additional 0·8 ml. for the volume of the ammonium sulphate precipitated). The solution (containing 1 g. of acetaldehyde in 100 ml.) is left to stand overnight, filtered and used for preparing standard solutions by diluting with alcohol. The solutions for comparison must be made up immediately before use as they cannot be kept, although the 1% solution can be stored.

BENZALDEHYDE

Methods of estimation

I. An estimation of benzaldehyde based on the preparation and isolation of the phenylhydrazone has been described as accurate and reliable although it is long and tedious.[1]

The solution containing benzaldehyde (not more than 0·5 g.) is made alkaline with sodium hydroxide (avoiding a large excess) and then extracted with ether (three lots of 25 ml. each). Freshly distilled phenylhydrazine (1·5 ml., see below) and an ethereal solution of glacial acetic acid (10 ml., 10%) are added to the combined ether extracts and the mixture is evaporated to

---

[1] G. A. Geiger, *J. Amer. Chem. Soc.* 1918, **40**, 1453–6; cf. M. Schubert and J. C. Dinkelspiel, *Ind. Eng. Chem.* (Anal. ed.), 1942, **14**, 154–5; W. Denis and P. B. Dunbar, *J. Ind. Eng. Chem.* 1909, **1**, 256–7; Assoc. Official Agr. Chem., *Official and Tentative Methods of Analysis*, 4th ed. 1935, p. 314, cf. pp. 136–44.

dryness, preferably in a gentle current of air. Water (50 ml.) is added to the residue, and after leaving to stand for 5–10 min. the insoluble benzaldehyde phenylhydrazone is filtered off into a weighed Gooch crucible and dried under reduced pressure at 70° C. for 2–3 hr.

PREPARATION: *Phenylhydrazine.* The phenylhydrazine either must be freshly distilled or have been stored in a refrigerator immediately after distillation for not more than 3 weeks.

II. The standard method of estimation with hydroxylamine hydrochloride (see pp. 144–52) has been modified by (a) replacing the expensive hydrochloride with the cheaper and more readily available sulphate, (b) changing the indicator to tetrabromophenol blue from bromophenol blue and (c) using aqueous sodium hydroxide instead of alcoholic sodium, or potassium, hydroxide for the titration. An accuracy of 0·6–0·7 % is claimed.[1]

III. The oxidation of benzaldehyde with excess chloramine-T and determination of this excess is claimed to give good results.[2]

A mixture of benzaldehyde (0·1–0·2 g.), potassium iodide (4–5 ml.) and chloramine-T (50 ml., 0·1 N) is made alkaline with aqueous sodium hydroxide. After allowing to stand for 30 min. the solution is acidified with dilute hydrochloric acid, and the liberated iodine is titrated with sodium thiosulphate (0·1 N).

SPECIFIC FOR PARTICULAR KETONES
ACETONE
Methods of estimation

I. Messinger's method

$$(CH_3)_2CO + 3I_2 + 4NaOH = CHI_3 + 3NaI + 3H_2O + CH_3.COONa.$$

Acetone reacts with iodine in the presence of alkali to form iodoform and acetic acid. An estimation based on this reaction has been investigated fully, and the conditions necessary for

[1] M. Schubert and J. C. Dinkelspiel, *loc. cit.* (p. 164, n. 1).
[2] B. Carli and R. Airoldi, *Ann. chim. applicata*, 1937, **27**, 56.

accurate and reproducible results have been determined.[1] An excess of iodine is added to acetone and an aqueous solution of sodium, or potassium, hydroxide, and the mixture is allowed to stand until the formation of iodoform is completed. The solution is acidified and the excess iodine is titrated with sodium thiosulphate. Reliable results are obtained when the following precautions are taken:

(i) The iodine is added at a uniform rate not exceeding 50 ml. in 3 min.

(ii) The flask is shaken continuously and vigorously during the addition of iodine.

(iii) The reaction mixture is left to stand for 15–20 min. before acidification, although fairly accurate results may be obtained after 5 min.

(iv) Sufficient acid is added to the alkaline solution to liberate the excess of iodine. Only a slight excess should be used, as low values result from titrations with sodium thiosulphate in strongly acid solution.

(v) The temperature at which the reaction mixture is kept is about 20° C.

(vi) A freshly prepared solution of starch is used as the indicator.

Iodine (50 ml., 0·1 N) is added slowly, with continuous shaking, to a mixture of aqueous solutions of acetone (25 ml., see below) and sodium, or potassium, hydroxide (50 ml., N) in a ground-glass stoppered bottle. After leaving to stand at about 20° C. for 15–20 min. the solution is acidified with sulphuric or hydrochloric acid (26 ml., 2 N), and the excess iodine is titrated with sodium thiosulphate (0·1 N). A freshly prepared solution of starch (1–2 ml., 1 %) is added just before the pale yellow colour of iodine disappears. It is essential to add starch as iodoform gives the solution a yellow colour. A blank titration is done under the same conditions as the estimation, and the difference

[1] J. Messinger, *Ber.* 1888, **21**, 3366–72; *J. Soc. Chem. Ind.* 1889, **18**, 138–9; F. Collischonn, *ibid.* 1891, **20**, 166–7; A. P. Sy, *J. Amer. Chem. Soc.* 1907, **29**, 786; L. F. Goodwin, *ibid.* 1920, **42**, 39–45; E. R. G. Ardagh, A. D. Barbour, G. E. McClellan and E. W. McBride, *J. Ind. Eng. Chem.* 1924, **16**, 1133–9; G. L. Stahly, O. L. Osburn and C. H. Werkman, *Analyst*, 1934, **59**, 319–25.

between the two titrations determines the amount of iodine which has been converted into iodoform.

The above method has been modified and adapted to the estimation of acetone in air.[1] A measured volume of air is drawn through a mixture of iodine (25 ml., 0·1 N) and sodium hydroxide (5 ml., 20%), the solution is acidified with hydrochloric acid (2 N) and the excess iodine is back-titrated with sodium thiosulphate (0·1 N) as above.

Acetone in factory gases can be estimated by this method.

PREPARATION: *Acetone.* Alternative methods for preparing the solution of acetone are:

(i) A glass ampoule (approximately $\frac{5}{8} \times \frac{3}{4}$ in.) is weighed before and after filling with acetone ($1·4 \pm 0·1$ g.) and then dropped into a glass-stoppered bottle (500 ml.) containing distilled water (200 ml.). The ampoule is broken and the capillary end is crushed with a glass rod to ensure that no acetone is trapped. The volume is then made up (to 1 l.) in a graduated flask with distilled water.

(ii) Acetone (1·75 ml., approximately 1·4 g.) is added from a pipette (2 ml., graduated in $\frac{1}{10}$ ml. divisions) to a small glass-stoppered weighing bottle. The bottle, which is weighed before and after adding the acetone, is inverted and the mouth is held under the surface of distilled water in a beaker (1 l.). The stopper is removed by a glass rod with a hook at one end. The bottle is washed thoroughly and the volume is made up as above.

II. The estimation with hydroxylamine hydrochloride has been tried and found to give low results owing to equilibrium being established when only 94·4% of the acetone has reacted.[2] A measured volume of air is drawn through two absorbers each having a column of hydroxylamine hydrochloride solution (0·2%) at least 40 cm. high at a rate not exceeding 20–25 l. an hour. The solutions are combined and titrated with sodium hydroxide (0·1 N) as described above (see p. 165).

If the hydroxylamine hydrochloride is standardized against a sample of pure acetone accurate results are obtained.

[1] Major Elliot and J. Dalton, *Analyst*, 1919, **44**, 132–6.
[2] M. Morasco, *J. Ind. Eng. Chem.* 1926, **18**, 701–2; cf. C. O. Haughton, *Ind. Eng. Chem.* (Anal. ed.), 1937, **9**, 167–8.

METHYL ETHYL KETONE

Methods of estimation

I. Reproducible and accurate results can be obtained by titration of the hydrochloric acid liberated in the oximation of the ketone with hydroxylamine hydrochloride (see p. 144).[1]

II. High values are obtained by the Messinger method (see p. 165) owing to a secondary reaction which involves 10 mol. of iodine for each mol. of the ketone instead of six as expected. The two reactions can be represented as follows:

$$CH_3.CH_2.CO.CH_3 + 3I_2 + 4NaOH = CHI_3 + 3NaI + CH_3CH_2.COONa + 3H_2O,$$
$$CH_3.CH_2.COCH_3 + 5I_2 + 5NaOH = 2CHI_3 + 4NaI + CH_3COONa + 4H_2O.$$

The reaction has been investigated, and a correction factor which allows accurate estimations to be made has been determined (100/107, 0·935).[2] The method is used commercially for the estimation of methyl ethyl ketone.

METHYL KETONES

Messinger's method can be used for the estimation of ketones which form iodoform. Correction factors which must be applied to allow for secondary reactions have been determined for a number of ketones:

Ketone	Correction factor
Methyl propyl ketone	100/106, 0·943
Methyl butyl ketone	100/97, 1·03
Methyl *iso*butyl ketone	100/104, 0·962

[1] Cf. A. A. Pryanishnikov, *Lesokhimicheskaya Pom.* 1933, no. 3, 27; M. Krajčinović, *Chem. Ztg.* 1931, **55**, 894–5.
[2] H. A. Cassar, *J. Ind. Eng. Chem.* 1927, **19**, 1061–2.

CHAPTER V header

CHAPTER V

AMINES AND AMINO COMPOUNDS

Amines and amino compounds vary so greatly in basicity and reactivity that no single method of estimation is ideal. The direct titration of these basic compounds with acid is only possible for the small number which are soluble in water and also sufficiently basic. The quantitative replacement of a mobile hydrogen of a primary or secondary base by acetylation, the condensation reaction between a primary amine or amino compound and benzaldehyde and the diazotization of a primary amino compound are reactions specific for each type of base. The methods of estimation, with one exception, use the characteristic reactions of the amino grouping. The indirect method involves the bromination of the aromatic nucleus, and it is the activating effect of the amino grouping which makes this reaction possible although the property is not specific for amino compounds. The following methods have been used for the estimation of primary, secondary and tertiary amines and amino compounds.

Acid titration:

$$R.NH_2 + HCl = R.NH_2.HCl.$$

Preparation and estimation of a reineckate:

$$R.NH_2 + NH_4^+[Cr(NH_3)_2(SCN)_4]^- H_2O \longrightarrow R.NH_3^+[Cr(NH_3)_2(SCN)_4]^- H_2O.$$

Acetylation:

A. With acetic anhydride:

$$R_2.NH + (CH_3.CO)_2O = R_2N.OC.CH_3 + CH_3.COOH.$$

B. With acetyl chloride:

$$R_2.NH + CH_3.COCl = R_2N.OC.CH_3 + HCl.$$

Preparation of a Schiff's base:

$$R.NH_2 + OCH.C_6H_5 = R.N{=}CH.C_6H_5 + H_2O.$$

Diazotization:

$$R.NH_2.HCl + ONOH = R.N_2Cl + 2H_2O.$$

Bromination:

$$\text{(aniline)} + 3Br_2 = \text{(2,4,6-tribromoaniline)} + 3HBr.$$

Reaction with picryl chloride:

$$R.NH_2 + Cl\text{(picryl)}NO_2 \xrightarrow{\text{NaHCO}_3} R.NH\text{(picryl)}NO_2 + NaCl + CO_2 + H_2O.$$

Miscellaneous.

ACID TITRATION

$$R.NH_2 + HCl = R.NH_2.HCl.$$

Many aliphatic amines are freely soluble in water, and their solutions are sufficiently basic for direct titration with a strong acid. This method is inapplicable to very weak bases, such as pyridine, and compounds which are insoluble or sparingly soluble in water. The latter group includes practically all amino compounds and a number of amines. These bases are not estimated usually by acid-alkali titration, although the following methods are useful occasionally.

(a) The base dissolved in alcohol is titrated with alcoholic hydrogen chloride using α-naphthyl red as the indicator.[1] Phenolphthalein cannot be used as amines, and amino compounds in 80 % alcohol do not react alkaline.[2]

(b) The hydrochloride is prepared, dissolved in water and the hot solution is titrated with a standard solution of sodium hydroxide (0·1 N) using phenolphthalein as the indicator.[3] The hydrochloride is made by adding hydrogen chloride in dry ether or benzene to the base dissolved in the same solvent and evaporating to dryness. It is convenient usually to add four-fifths of the sodium hydroxide before heating the mixture and carrying out the titration. The method is unsatisfactory for

[1] Cf. K. Linderström-Lang, *Z. physiol. Chem.* 1928, **173**, 32–50.
[2] F. W. Foreman, *Biochem. J.* 1920, **14**, 451–73.
[3] B. P. Fedorov and A. A. Spryskov, *Org. Chem. Ind.* (U.S.S.R.), 1936, **1**, 620.

p-aminophenol, quinoline, pyridine, α-aminoanthraquinone and dimethylaniline.

PREPARATION: *α-Naphthyl red* (benzeneazo-α-naphthylamine). α-Naphthyl red (0·1 g.) is dissolved in alcohol (100 c.c., 96 %).

PREPARATION AND ESTIMATION OF A REINECKATE

$$R.NH_2 + NH_4^+[Cr(NH_3)_2(SCN)_4]^-H_2O \longrightarrow R.NH_3^+[Cr(NH_3)_2(SCN)_4]^-H_2O.$$

Reinecke's salt, prepared from potassium dichromate and ammonium thiocyanate, reacts with amines and amino compounds to form complexes in which 1 mol. of the base replaces one ammonium group in the salt. The solubility in water of the complexes depends on the amine or amino compound, and in general it is lowest for tertiary bases and highest for primary amines.[1] A gravimetric method of analysis based on the above reaction has been used for the estimation of amines, amino compounds, and quaternary ammonium salts.[2]

The complex must be precipitated under controlled conditions owing to the ease with which colloidal precipitates which are extremely difficult to filter, can be formed. The conditions given below for the formation and precipitation of the insoluble complex are usually satisfactory, but occasionally it is necessary to vary them slightly. Heterocyclic amines react in strongly acid solution, and creatine and amino alcohols are best precipitated in weakly acid media. A limited number of amines, especially primary amines, form reineckates which are not sufficiently insoluble for satisfactory gravimetric analysis. The majority of the alkaloids can be estimated by this method although precipitation is not always quantitative.[3]

An ice-cold solution of Reinecke's salt (a large excess) is added to a cold, neutral or acid, solution of the base (as concentrated as possible). A rose-coloured precipitate should be formed,

[1] O. T. Christensen, *J. prakt. chem.* 1892, **45**, 213–22, 256–76.

[2] A. Dansi, L. Mamoli and B. Ciocca, *Ann. chim. applic.* 1932, **22**, 561–5; A. Hantzsch and H. Carlsohn, *Z. anorg. chem.* 1926, **156**, 199–209; F. Hein and F. A. Segitz, *ibid.* 1926, **158**, 153–74; 1927, **72**, 119–21.

[3] L. Rosenthaler, *Arch. Pharm.* 1927, **265**, 319–23; A. Niethammer, *Biochem. Z.* 1929, **213**, 138–41; E. J. Kahane, *J. Pharm. chim.* 1935, **22**, 254–67; P. Duquenois and Mlle Faller, *Bull. soc. chim.* 1939, **6**, 998–1008.

although it may be necessary to acidify the mixture (to congo red). As the solubility in water of the complexes increases rapidly with rise in temperature the solution should be kept as cold as possible for 24 hr. and then filtered (see below). The solid is washed successively with ice-cold water, alcohol and ether and dried either at room temperature under reduced pressure or at 65–70° C. The complex can be weighed, or it can be ignited and weighed as chromic oxide or the percentage of nitrogen can be determined by combustion. If the complex has been freed from unchanged salt seven atoms of nitrogen are present to every one in the base.

If a correction of 0·5 part per 100 is made the complex can be filtered off after standing for 30 min.

Accuracy ± 0·5 %.

PREPARATIONS: *Reinecke's salt*. The salt can be prepared as a monohydrate or in the anhydrous condition by varying the method of preparation, and it is the former which is used.[1]

Ammonium thiocyanate (200 g.) is heated to 140° C., and potassium dichromate (40 g.) is added in successive portions with shaking so that the reaction temperature is maintained between 140 and 160° C. A red mass is obtained on cooling which is triturated with water (250 ml.) and then washed with cold water to remove excess of either reactant. The residue is extracted repeatedly with warm water (50–60° C.) until the filtrate no longer deposits solid on cooling. The combined extracts are then added to a saturated solution of ammonium chloride. The hydrated salt only is precipitated under these conditions, and it is filtered off, washed with cold water, and dried in a vacuum desiccator.

The salt decomposes slowly in light but can be kept in the dark. The solution decomposes in a few days in the cold and rapidly on heating. Decomposition can be detected by the thiocyanate reaction with a ferric salt.

Saturated solution of Reinecke's salt. The salt is dissolved in hot water (40 % solution), and the solution cooled and filtered after several hours.

[1] J. Morland, *J. Chem. Soc.* 1861, **13**, 252–4; O. Reinecke, *Ann.* 1863, **126**, 113–18.

ACETYLATION

A. With Acetic Anhydride

$$R_2.NH + (CH_3.CO)_2O = R_2N.OC.CH_3 + CH_3.COOH.$$

The Karl Fischer reagent can be used for the indirect estimation of primary and secondary amines and amino compounds.[1] An excess of acetic anhydride and pyridine is added to the base, and after acetylation is finished the unchanged anhydride is hydrolysed with a known volume of water. The mixture is then titrated with the Karl Fischer reagent.[2]

The method can be used for primary and secondary amino compounds and aliphatic, alicyclic and heterocyclic amines. Tertiary bases do not react with acetic anhydride, but they can be estimated indirectly in a mixture of bases if the total basicity is found by another method (see p. 170, 183). Amides, urethanes, nitriles and tertiary amines and amino compounds usually have no effect. When a primary alcoholic group also is present, as in the ethanolamines, the procedure is modified to effect the quantitative acetylation of both the amino and alcoholic groupings. Di-s-butylamine, diphenylamine, carbazole, pyrrole, phenothiazine and diaryl secondary amino compounds cannot be estimated by this method.

The acetylating agent (20 ml.) is added to the primary or secondary base, or to a mixture of bases (up to 10 m.equiv.), and weighed into a glass-stoppered flask (250 ml.). A blank is prepared and the two flasks are stoppered tightly, shaken and left to stand for 30 min. At the end of this time the hydrolysing agent (25 ml.) is added. The flasks are placed in a water-bath at $60 \pm 1°$ C., and after taking the stoppers out momentarily, to allow for the expansion of the air, they are sealed tightly (if possible the stoppers should be held by spring clips) and heated for 30 min. The flasks are taken from the water-bath, cooled to room temperature and the contents are titrated with the Karl Fischer reagent. The free water in the sample is found by

---

[1] J. Mitchell, Jr., W. Hawkins and D. M. Smith, *J. Amer. Chem. Soc.* 1944, **66**, 782–4.

[2] Cf. 'Estimation of anhydrides by the Karl Fischer reagent', D. M. Smith, W. M. D. Bryant and J. Mitchell, Jr., *ibid.* 1941, **63**, 1700–1.

titrating a solution of the base in glacial acetic acid with Karl Fischer reagent (see p. 51).

When a primary alcoholic group is present the mixture is heated longer (for 1 hr.). Secondary alcohols are not esterified quantitatively under these conditions and tertiary alcohols react only slightly.

PREPARATIONS: *Karl Fischer reagent.* See p. 55.

Acetylating mixture. Pure acetic anhydride (142 ml., 1·5 mol.) is added to dry pyridine (800 ml.) and the mixture made up (to 1 l.) with pyridine.

Hydrolysing agent. Sodium iodide (100 g., A.R.) is added to distilled water (22 ml.) and the solution is made up (to 1 l.) with dry pyridine.

B. With Acetyl Chloride

$$R_2.NH + Cl.COCH_3 = R_2N.OC.CH_3 + HCl.$$

Although alcohols and phenols can be estimated satisfactorily using acetyl pyridinium chloride in toluene, an identical procedure cannot be applied to amines and amino compounds (cf. p. 47). Quantitative results can be obtained sometimes using the same acetylating agent in di-n-butyl ether as the solvent instead of toluene and by raising the temperature of the reaction from 60 to 70° C.[1] Acetylation values for many primary and secondary amino compounds are over 90 %, but the method is not completely reliable as results as low, sometimes, as 13 % of the theoretical are obtained. This is generally so when the base is very insoluble in the acetylating mixture as shown by tribromoaniline and p-nitroacetanilide. A limited number of compounds only have been estimated by this method and of these aniline and o-nitroaniline give very low results.

The procedure used for alcohols and phenols (p. 47) is modified by changing the acetylating agent and raising the temperature of the reaction to 70° C., and by using a long neck round-bottomed flask (300 ml.).

PREPARATION: *Acetyl chloride solution.* A solution of acetyl chloride (1·5 M) in dry, peroxide-free di-n-butyl ether is prepared

[1] V. R. Olson and H. B. Feldman, *J. Amer. Chem. Soc.* 1937, **59**, 2003–5.

and samples are withdrawn and used in the above estimation. The ether must be free from peroxides and can be prepared according to the method of Brandt.[1]

PREPARATION OF A SCHIFF'S BASE

$$R.NH_2 + OCH.C_6H_5 = R.N{=}CH.C_6H_5 + H_2O.$$

The reaction of an amine or amino compound and an aldehyde takes place readily, forming water and a weakly basic compound known as a Schiff's base. The products formed by condensation of aliphatic aldehydes, especially formaldehyde, and primary bases are often complex owing to polymerization of the initial product, but aromatic aldehydes under the same conditions yield stable derivatives. The condensation has been adapted to the estimation of primary amines and amino compounds by using an excess of benzaldehyde to force the reaction to completion and titrating the water formed with the Karl Fischer reagent.[2] Secondary and tertiary amines and amino compounds do not interfere appreciably and the method can be used for the estimation of primary bases in a mixture. As the primary and secondary base can be determined by acetylation, and as the total basicity can be estimated directly by titration with an acid, the different amounts of the amines and amino compounds can be found by a suitable combination of methods.

Primary bases in the aliphatic, aromatic and alicyclic series and amino alcohols can be estimated by this method. The reaction of the last class of compounds may be either a condensation involving both the amino and hydroxyl groups to form a substituted oxazine, or the normal reaction to give an imine.[3] Only one equivalent of water is formed by either reaction, and analysis therefore is unaffected. Amino alcohols containing a secondary amino group react almost quantitatively (e.g. diethanolamine gives 1 mol. of water for each mol. of the amine, and di*iso*propylamine and hydroxyethylethylenediamine give 0·9 and 1·8 respectively). Heterocyclic bases interfere with the

[1] L. Brandt, *Chem.-Ztg.* 1927, **51**, 891–3.
[2] W. Hawkins, D. M. Smith and J. Mitchell, Jr., *J. Amer. Chem. Soc.* 1944, **66**, 1662–3.
[3] Cf. M. Kohn, *Monatsh.* 1905, **26**, 95–8.

estimation as they react with benzaldehyde eliminating one mol. of water for every 2 mol. of the amine to give NN'-benzaldimines. Piperazine and morpholine appear to react in the same way. Results consistently low (5%) have been recorded with p-bromoaniline. Urea and methyl urea only react to a limited degree (10 and 15% respectively). In the original experiments negative results were obtained with di*iso*butylamine, methylaniline, diphenylamine, carbazole, triethanolamine, and tri*iso*propanolamine. Diethyl- and di-n-butyl-amine gave a 1–2% reaction, but impurities may have been present in the samples tested.

The benzaldehyde used in the condensation must be free from benzoic acid (see below) as low values are always obtained when the acid is present (10% of benzoic acid in the aldehyde gives results up to 5% low). The excess of aldehyde is converted into the cyanohydrin before the water formed in the reaction is titrated with the Karl Fischer reagent. A blank determination should always be carried out to estimate the free water in the benzaldehyde and pyridine. The titration of the reaction mixture must be done fairly quickly and the first end-point taken as fading occurs occasionally on standing for a few minutes.

The primary amine or amino compound (0·1M) is weighed and added to a graduated flask filled about one-third full with dry pyridine, and a standard solution is made up by adding more solvent. Aliquot portions (10 ml.) are transferred to a stoppered bottle (250 ml.), benzaldehyde (3 ml.) is added, the bottle is stoppered and placed in a thermostatically controlled water-bath ($60 \pm 1°$) for half an hour. The mixture is allowed to cool to room temperature, moved into a fume cupboard and sodium cyanide (0·2 g.) and a solution of hydrogen cyanide (30 ml., 6%) in dry pyridine added. The sodium cyanide catalyses the reaction, and as it is practically insoluble in pyridine the bottle must be shaken vigorously for 1 min. to start the cyanohydrin formation. The mixture is allowed to stand for 45 min. and then titrated rapidly with the Karl Fischer reagent (see p. 51). The first sharp end-point is taken.

A blank determination must be carried out under identical conditions, and if the amine is not anhydrous the percentage of water must be determined by titrating an acetic acid solution of the original sample with the Karl Fischer reagent.

Percentage of —NH$_2$

$$= \frac{\text{ml. of reagent (sample} - \text{blank)} \times \text{normality} \times 16}{100 \times \text{weight of amine}}.$$

Accuracy $\pm 0.2\%$.

PREPARATION: *Benzaldehyde.* The benzaldehyde is redistilled immediately before use or oxidation is inhibited by the addition of hydroquinone (0·1%) after distillation.

DIAZOTIZATION

$$R.NH_2.HCl + ONOH = R.N_2Cl + 2H_2O.$$

The quantitative diazotization of a primary amino compound has been used frequently as a means of estimation, but the direct titration with a standard solution of sodium nitrite is accurate only for a limited number of bases.[1] When diazotization takes place slowly the reaction is incomplete and low values are obtained, and when the time of reaction is increased appreciably nitrous acid is lost and the results are high and variable. A procedure which is satisfactory for most amino compounds is possible by dissolving the base in hydrochloric acid, adding an excess of sodium nitrite solution and back-titrating with a standard solution of a primary amino which is known to diazotize rapidly.[2] *p*-Nitroaniline or nitrocresidine are suitable bases, as they react rapidly and are also sufficiently stable to be used for preparing standard solutions. The loss of nitrous acid is prevented by adding nitric acid to the reaction mixture immediately after the solution of sodium nitrite:

$$3HNO_2 \rightleftharpoons HNO_3 + 2NO + H_2O.$$

The end-point is found either by spotting on starch iodide paper as an external indicator or potentiometrically using a platinum electrode and a standard calomel cell as the reference electrode.[3]

[1] G. Lunge and E. Berl, *Chemische-technische Untersuchungsmethoden*; Th. Sabalitschka and H. Schrader, *Z. anal. Chem.* 1921, **34**, 45.

[2] H. R. Lee and D. C. Jones, *Ind. Eng. Chem.* 1924, **16**, 930–31; *ibid.* 1924, **16**, 948–9.

[3] E. Müller and E. Dachselt, *Z. Electrochem.* 1926, **31**, 662–6; B. Singh and G. Ahmad, *J. Indian Chem. Soc.* 1938, **15**, 416–20; B. Singh and A. Rehmann, *ibid.* 1942, **19**, 349–53.

The technique necessary for the first of these methods may be difficult to acquire, but a few titrations usually give the experience necessary to obtain accurate values. The blue colour develops slowly, but the speed at which it appears has no relation to the end-point and at the end-point the blue colour appears rather slowly (about 2 sec.). Aniline, the toluidines, xylidenes, dehydrothioaminosulphonic acids, primuline and similar bases have been estimated successfully in this way. The potentiometric method has been used for aniline, the toluidines, m-xylidene, o- and p-aminobenzoic acid, p-aminochlorobenzene, p-aminobenzenesulphonic acid, $1:2:4$-nitroethoxyaniline, $1:2:4$-aminonitrophenol, sodium azobenzenedisulphonamide. o-Phenylenediamine reacts quantitatively with nitrous acid to give a benztriazole, and the curve rises and falls slightly before rising sharply at the end-point. A modification of the method has been described in which the amino compound is diazotized in the normal way and then heated to give off an equivalent of nitrogen.[1] The values so obtained are less accurate and the estimation is more cumbersome and complicated than titration methods.[2]

The amino compound ($\frac{1}{25}$ mol.) is dissolved in hydrochloric acid (5 %) and made up to 500 ml. with acid of the same strength. A measured volume of this solution (50 ml.) is placed in a beaker and diluted (to 300 ml.) with water. Concentrated hydrochloric acid (25 ml.) is added, the liquid cooled in an ice bath (to between 0–5° C.) and a little crushed ice placed in the beaker. An aqueous solution of sodium nitrite (50 ml., 0·1 N) is added with continuous stirring followed immediately by concentrated nitric acid free from nitrogen trioxide (10 ml.). The beaker is covered with a watch-glass and the mixture is left to stand in the ice-bath, with occasional stirring, for 30 min. The excess sodium nitrite is determined by titration with either p-nitroaniline or nitrocresidien, using an external indicator (starch iodide paper) or potentiometrically.

A standard calomel cell is used as the reference electrode in

[1] E. Grigoriev, Z. anal. Chem. 1926, **69**, 47–50.

[2] A. V. Pamfilov, Ind. Eng. Chem. 1926, **18**, 763–4; Z. anal. Chem. 1926, **69**, 282–92.

the potentiometric titration. The solution is stirred for 2–5 min. after each addition of sodium nitrite solution, and a micro-burette should be used near the end-point. The slope of the curve at 0·54–0·59 volt makes it possible to use an indicator electrode at this potential until the galvanometer gives a permanent zero reading. Below 20° C. the results are reliable, but above this temperature the accuracy falls off rapidly.

The amount of nitric acid which must be added to stop the decomposition of the nitrous acid depends upon the amount of sodium nitrite which is present in excess. The above conditions, using 10 ml. of concentrated nitric acid with an excess of 10 ml. of sodium nitrite, are regarded as the most suitable.

Accuracy $\pm 0\cdot1\%$.

PREPARATION: *p-Nitroaniline.* *p*-Nitroaniline (14·0 g.) is dissolved in hot water (200 ml.) and concentrated hydrochloric acid (150 ml.). After standing overnight the solution is filtered and made up (to 1 l.) in a graduated flask. The solution of sodium nitrite is standardized against sulphanilic acid of known purity and then used to titrate the *p*-nitroaniline. The *p*-nitro-aniline (10 ml.) is made up to the same volume (400 ml.) with water and titrated under the same conditions as described above for the back-titration.

The solution of *p*-nitroaniline can be made up and stored for a fortnight at least without any appreciable deterioration.

BROMINATION

Amino compounds brominate very easily and they can be estimated by bromine titration as described under phenols (see pp. 86–90).

Some amino compounds, such as *o*- and *p*-toluidine, are oxidized as well as brominated with bromine water, and abnormal and indefinite values are obtained when an excess of the reagent is used. Good results sometimes can be found by direct titration, although this method is long and tedious (by direct titration the values found for *o*- and *p*-toluidine are 100·08 and 100·19 % respectively). Aniline and compounds which yield aniline on hydrolysis or form tribromoaniline by replacement of a grouping (e.g. sulphanilic and anthranilic acids, see p. 87) also are sensitive to oxidizing agents, but they can be estimated by indirect titration if the reaction is done at 10–15° C. This method cannot be used for *p*-aminobenzoic acid, as the insoluble bromo derivative is a mixture, probably of the dibromo and tribromo derivatives. A low value is always obtained with *p*-nitroaniline, probably owing to occlusion in the very bulky precipitate which is formed. Acetanilide is estimated by hydrolysing with boiling hydrochloric acid (6N) for 10–15 min., neutralizing the reaction mixture with sodium hydroxide, making slightly acid with hydrochloric acid, diluting the solution (to 250 ml.) and then carrying out the estimation in the usual way.

The amino compound is dissolved either in dilute hydrochloric acid or less frequently in dilute sodium hydroxide (e.g. anthranilic acid). The estimation is as described under phenols (see pp. 86–90).

Amino compound	Time of bromination in min.
Aniline	5–10
p-Chloroaniline	10
o-Nitroaniline	30
m-Nitroaniline	30
m-Aminobenzoic acid	10–15
m-Toluidine	5–10
Anthranilic acid	30
Sulphanilic acid	30
Metanilic acid	5–15
Acetanilide	5–10

REACTION WITH PICRYL CHLORIDE

$$R.NH_2 + Cl \underset{NO_2}{\overset{NO_2}{\bigcirc}} NO_2 \xrightarrow{NaHCO_3} R.NH \underset{NO_2}{\overset{NO_2}{\bigcirc}} NO_2 + NaCl + CO_2 + H_2O.$$

The reaction between picryl chloride (2-chloro-1:3:5-trinitro-benzene) and an amino compound can be used for the estimation of a limited number of such bases.[1] The amino compound is dissolved in ethyl acetate and added to a slight excess of picryl chloride solution in the same solvent and a little sodium bicarbonate. Sodium chloride is formed under these conditions and can be extracted from the mixture with water and estimated either volumetrically or gravimetrically.

Aniline, o- and p-toluidines and o- and p-anisidine react quantitatively, but when a negative group is present either the reaction does not take place at all, as with o-nitroaniline, or only incompletely, as with anthranilic acid. The reaction with α-naphthylamine is almost quantitative, but the β-compound behaves irregularly. Both amino groups in benzidine react, but the extraction and estimation of the sodium chloride is difficult. Diamino compounds do not combine quantitatively, and secondary bases hardly react at all.

The base (approximately 0·1 M) in ethyl acetate is added to a strong picryl chloride solution (slight excess) in ethyl acetate and sodium bicarbonate (0·1 g.). The mixture is left to stand at room temperature (15 min. for aniline, 1 hr. for all other amino compounds), water (250 ml.) is added and the mixture heated for 15 min. on a water-bath. After leaving to stand at room temperature for a further 3 hr. the sodium chloride is filtered off and the chlorine is estimated either gravimetrically or volumetrically. It is unnecessary to filter off the sodium chloride if Mohr's titration method is used.

The method has been modified and used for the estimation of aniline in the presence of alkylanilines.[2] Alkylanilines react with picryl chloride, but only very slowly, and under the conditions of the estimation this reaction can be neglected.[3] The method is a rapid and satisfactory means of determining varying amounts of aniline in some of the commoner alkylanilines.

The sample (10 g.) containing aniline is weighed into a beaker, and benzene (30 ml.) and picryl chloride in benzene (10 ml., 10 %) are added with stirring. After standing a few minutes at

[1] B. Linke, H. Preissecker and J. Stadler, Ber. 1925, 65, 1280–4.

[2] G. Spencer and J. E. Brimley, J. Soc. Chem. Ind. 1945, 64, 53–5; A. Néljubina, Anilinokras. Prom. 1924, 4, 120.

[3] A. H. Rheinlander, J. Chem. Soc. 1923, 123, 3099–110.

room temperature the mixture is transferred to a separating funnel and extracted five times with water (50 ml. each time). The combined extracts are titrated with a standard solution of sodium hydroxide (0·1 N) using phenolphthalein as the indicator.

The percentage of aniline in the sample should not exceed 20 % when methyl- or ethyl-aniline are the alkylanilines, 10 % for diethylaniline and about 2 % for dimethyl- or benzylethyl-aniline. For larger amounts a sample is weighed equivalent to between 20 and 40 ml. of a 0·1 N solution of sodium hydroxide, and diethylaniline (10 ml.) or dimethylaniline (30 ml.) is added and the estimation is then completed as above. A blank determination should be carried out to ensure that the added alkylaniline is neutral.

MISCELLANEOUS

(1) Long-chain aliphatic amines, such as dioctylmethylamine, can be estimated colorimetrically by extracting into light petroleum with dilute acid and adding chloroform and picric acid. The colour which develops is measured in a Spekker photoelectric absorptiometer using a no. 7 dark blue filter.[1]

(2) Low concentrations of primary amino compounds sometimes can be estimated colorimetrically by diazotizing and coupling with α- or β-naphthol in alkaline solution.[2] Sodium nitrite (0·2 ml., 1 %) and hydrochloric acid (1 ml., N) are added to a solution of the base (about 0·2 mg.) and the mixture is shaken. After standing for a few minutes an acetone solution of β-naphthol (1 ml.) and sodium hydroxide (2 ml., N) are added, the mixture is shaken until a red colour develops. On standing two layers form. The top one is transferred to a graduated flask (25 ml.) and made up to the mark with distilled water. The solution is mixed and the colour is compared either with a standard solution or the colour density is measured in a calibrated instrument. The red colour reaches its maximum intensity after 3 min. and does not change for an hour.

[1] B. M. C. Hopewell and J. E. Page, *Analyst*, 1945, **70**, 17.

[2] H. Seydlitz, *Svensk Farm. Tid.* 1946, **50**, 65–70; E. Havinga, *Rec. trav. chim.* 1944, **63**, 243–7; P. Fantl, *Australian J. Exp. Biol. Med. Sci.* 1940, **18**, 175–84; J. Haslam and A. H. S. Guthrie, *Analyst*, 1943, **68**, 328–30.

The method is suitable only for those primary amino compounds which can be diazotized easily (cf. p. 177).

Accuracy $\pm 2\%$.

(3) An aqueous solution of perchloric acid containing acetic anhydride equivalent to the amount of water present may be used for the direct titration of primary, secondary and tertiary bases. The solution is made up to the desired volume with glacial acetic acid. The end-point is found with indicators or potentiometrically.[1]

(4) Primary amino compounds dissolved in dry benzene or toluene can be titrated directly with acetic anhydride. Water interferes with the estimation, and the amount present must not exceed a few millilitres. The reaction is completed when a drop of the reaction mixture fails to give a yellow spot on a ground pulp paper (newsprint).[2]

(5) Primary and secondary bases can be estimated using an acetylating mixture of acetic anhydride (1 vol.) and pyridine (4 vol.). The sample is heated with at least twice the theoretical amount of the reagent and then titrated with a standard solution of alcoholic sodium hydroxide. A blank determination is carried out in parallel.[3]

(6) A specific method for the estimation of primary amines is based on the following reaction with nitrous acid:

$$R.NH_2 + ONOH = R.OH + H_2O + N_2.$$

A large volume of nitric oxide is liberated in the reaction, and the nitrogen is extracted by transferring the mixture of gases to a modified Hempel pipette and shaking with potassium permanganate to remove the nitric oxide. In the absence of ammonia and urea samples can be analysed satisfactorily in the standard Van Slyke apparatus if they contain not less than 0.0005 mg. and not more than 0.50 mg. of amino nitrogen.[4]

[1] K. Blumrich and G. Bandel, *Angew. Chem.* 1941, **54**, 374–5.
[2] V. N. Ivanov, *Org. Chem. Ind.* (U.S.S.R.), 1937, **3**, 162–3.
[3] F. H. Stodola, *Mikrochemie*, 1937, **21**, 180–3; cf. p. 43.
[4] D. D. Van Slyke, *J. Biol. Chem.* 1911, **9**, 195 *et seq.* to 1915, **23**, 407.

CHAPTER VI

NITRO, CYANO AND NITROSO COMPOUNDS, ISOCYANATES AND ISOTHIOCYANATES

NITRO COMPOUNDS

A limited number of methods only are used for the estimation of nitro compounds, and of these reduction to the corresponding primary base is employed most often. Several methods have been described using different reducing agents, of which titanous and stannous chlorides are usually the most satisfactory. Kjeldahl's method of estimation can be modified and used, and colorimetric methods have been described for small concentrations of aliphatic or aromatic nitro compounds.

Reduction:

  A. To the corresponding primary base:

$$R.NO_2 + 6H = R.NH_2 + 2H_2O.$$

  B. To ammonia.

Colorimetric:

  A. Specific for mononitroparaffins.

  B. Specific for aromatic nitro compounds.

Miscellaneous.

REDUCTION

A. To the Corresponding Primary Base

$$R.NO_2 + 6H = R.NH_2 + 2H_2O.$$

The formation of an amino compound by reduction of an aromatic nitro group is used frequently as a method of determining the percentage of nitro group present. The amino compound can be titrated directly (see p. 170), or, more generally, the excess of the reducing agent is estimated. Stannous chloride was used originally,[1] and many modifications of the method have been proposed since by many investigators.[2]

The nitro compound is usually reduced completely by heating in an inert atmosphere with excess of the reducing agent dissolved in hydrochloric acid. The solution is cooled, and the excess of stannous chloride is titrated either with iodine or with ferric chloride using starch and ammonium thiocyanate respectively as the indicator. The more volatile nitro compounds should be sulphonated before reduction, and those which are insoluble in water can either be sulphonated or dissolved in a small volume of alcohol.

The method has been found to give low results with many substances, and originally this was thought to be due to a loss of volatile nitro compounds on boiling, or to incomplete reduction. These explanations now are known to be inadequate as low values are still found with some substances when the reduction is carried out in sealed tubes (e.g. only two-thirds of the nitrogen is estimated in p-nitrochlorobenzene). It has been shown that chlorination can take place simultaneously with reduction when nitro compounds are reduced in the presence of either hydrochloric acid, or the chloride ion.[3] Chlorination liberates one atom of hydrogen for each atom of chlorine substituted and can therefore reduce the amount of stannous chloride required by one-third. It has been stated, although it is not acknowledged universally, that chlorination is more pronounced in alcoholic solution.

Titanous chloride has been proposed as an alternative

[1] H. Limpricht. *Ber.* 1878, **11**, 35–42.

[2] S. W. Young and R. E. Swain, *J. Amer. Chem. Soc.* 1897, **19**, 812–14; A. J. Berry and C. K. Colwell, *Chem. News*, 1915, **112**, 1–2; P. Altmann, *J. prakt. chem.* 1901, [ii], **63**, 370–80; *J. Soc. Chem. Ind.* 1901, **20**, 622–3; A. P. Sachs, *ibid.* 1917, **36**, 915–16; E. de W. S. Colver and E. B. R. Prideaux, *ibid.* 1917, **36**, 480–3; L. Desvergnes, *Ann. Chim. Analyst*, 1920, **2**, 141–3; D. Florentin and H. Vandenberghe, *Bull. Soc. Chim.* 1920, **27**, 158–66; G. Wallerius, *Tek. Tid., Uppl. C* (Kemi), 1928, **58**, 33–5; L. E. Hinkel, E. E. Ayling and T. M. Walters, *J. Chem. Soc.* 1939, pp. 403–6.

[3] T. Callan, J. A. R. Henderson and N. Strafford, *J. Soc. Chem. Ind.* 1920, **39**, 86–88T; R. Fittig, *Ber.* 1875, **8**, 15–16; P. Seidler, *ibid.* 1878, **11**, 1201–2; E. Kock, *ibid.* 1887, **20**, 1567–9; E. Bamberger, *ibid.* 1895, **28**, 245–51; S. Gabriel and R. Stelzner, *ibid.* 1896, **29**, 303–10; W. G. Hurst and J. F. Thorpe, *J. Chem. Soc.* 1915, **107**, 934–41.

reducing agent to stannous chloride.[1] This change of reagent does not reduce the possibility of chlorination, and for this reason a lowering of the chloride-ion concentration has been advocated either by substituting sulphuric for hydrochloric acid, or by using titanous sulphate.[2] The excess titanous salt is estimated by titrating with ferric alum using ammonium thiocyanate as the indicator.

The method has been criticized adversely and inaccuracies have been reported, but when standard conditions are used and a blank estimation is done reliable values are obtained. More consistent and accurate results can be obtained by standardizing the titanous salt against an organic nitro compound instead of an inorganic salt.[3] Pure, recrystallized, p-nitroaniline,[4] 3-nitro-p-toluidine and m-nitroaniline have been recommended.[5]

The nitro compound can be estimated by titration and the end-point determined either using an indicator or electrometrically.[6]

1. *Titration with a titanous salt*

$$6TiCl_3 + 6HCl + R.NO_2 = 6TiCl_4 + R.NH_2 + 2H_2O,$$
$$TiCl_3 + FeCl_3 = TiCl_4 + FeCl_2.$$

(*a*) Nitro compounds soluble in water. An aqueous solution of the nitro compound (0·5–2·0 g.) is acidified with sulphuric acid (25 ml., 40%) and a slow stream of carbon dioxide (1–2 bubbles a second) is passed through as the liquid is heated to boiling. The heating must be done under an efficient reflux condenser in a flask fitted with a ground-glass joint if the nitro compound is volatile in steam. Heating is stopped momentarily, titanous

[1] E. Knecht, *Ber.* 1903, **36**, 166–9; E. Knecht and E. Hibbert, *ibid.* 1903, **36**, 1549–55; *ibid.* 1905, **38**, 3318–26; H. Rathsburg, *ibid.* 1921, **54**B, 3183–4; C. F. van Duin, *Rec. trav. chim.* 1920, **39**, 578–85; T. Callan, J. A. R. Henderson and N. Strafford, *loc. cit.* (p. 185, n. 3); F. L. English, *J. Ind. Eng. Chem.* 1920, **12**, 994–7; W. W. Becker, *Ind. Eng. Chem.* (Anal. ed.), 1933, **5**, 152–4; T. Callan and J. A. R. Henderson, *J. Soc. Chem. Ind.* 1922, **41**, 157–61T; T. Callan and N. Strafford, *ibid.* 1924, **43**, 1–8T; S. Mayuramen, *Sci. Papers Inst. Phys-Chem. Research* (Tokyo), 1931, **16**, 196.

[2] C. F. van Duin; H. Rathsburg; T. Callan and J. A. R. Henderson; W. W. Becker, *loc. cit.* (p. 186, n. 1).

[3] T. Callan and J. A. R. Henderson, *loc. cit.* (p. 186, n. 1).

[4] *Ibid.*

[5] F. L. English, *loc. cit.* (p. 186, n. 1).

[6] I. M. Kolthoff and C. Robinson, *Rec. trav. chim.* 1926, **45**, 169–76; E. Dachselt, *Z. anal. chem.* 1926, **68**, 404–10.

sulphate or chloride (50 ml., 0·2 N, an excess of about 10 ml.) is added and the reduction finished in an atmosphere of carbon dioxide by heating during a further 10 min. The solution is cooled, and the excess of the titanous salt is found by titration with a standard solution of ferric alum (0·3 N) using potassium thiocyanate (10 ml., 10 %) as the indicator. A blank determination is done under identical conditions, and the difference between the two readings is a measure of the titanous salt used in the reduction of the nitro group.

(b) Nitro compounds insoluble in water. The nitro compound (1 part) and fuming sulphuric acid (20 parts by weight) are heated on a water bath for 2 hr. The solution of the sulphonated nitro compound is cautiously made up to a known volume with water and then estimated as above.

If the nitro compound is insoluble and not sulphonated easily it is dissolved in alcohol, and this solution added to a known volume of the acidified titanous salt. The mixture is boiled, allowed to cool, and the excess of the reagent estimated as above. Carbon dioxide is passed through the solution of the titanous salt during the estimation.

(c) Electrometric titration. A known volume of the titanous salt is added to a solution of the nitro compound and either Rochelle salt or preferably sodium citrate until a distinct violet colour persists (an excess of 1–3 ml.). The mixture is titrated with ferric chloride and at the end-point a sharp change in potential takes place (cf. pp. 195–6).

It is claimed that in the presence of Rochelle salt, or sodium citrate, direct titration is possible, as a sharp colour change to violet takes place on the addition of an excess (1 drop) of ferric chloride.[1]

2. *Titration with stannous chloride*

$$R.NO_2 + 3SnCl_2 + 6HCl = R.NH_2 + 3SnCl_4 + 2H_2O.$$

The nitro compound (0·1–0·2 g.), dissolved in aldehyde-free alcohol (5–6 ml., see below) if insoluble in water, is placed in a flask and a slow stream of dry carbon dioxide (1–2 bubbles a second) is bubbled through the solution during the estimation.

[1] I. M. Kolthoff and C. Robinson, *loc. cit.* (p. 186, n. 6).

Sulphuric acid (10 ml., 1:1 by volume) and a standard solution of stannous chloride (10 ml., see below) are added and the mixture is heated gradually to 100° C. and kept at this temperature during 1–5 hr. The mixture is cooled, diluted to 200 ml. with water saturated with carbon dioxide, and the excess of stannous chloride is titrated with a standard solution of iodine (approximately 0·6 N) using starch as the indicator.

A blank estimation should be done.

PREPARATIONS: *Standard solution of stannous chloride.* Stannous chloride (283 g., A.R.) is dissolved in hydrochloric acid (300 ml., $d = 1·16$) and the solution made up (to 1 l.) with water saturated with carbon dioxide. The solution is approximately 2·5 N in 3 N hydrochloric acid.

The estimation is affected by the purity of the tin used in the preparation of the stannous chloride, but the effect is eliminated by using pure tin and doing a blank titration.

Alcohol free from aldehyde. A rapid stream of carbon dioxide is passed during several hours through the alcohol heated under a reflux condenser (cf. p. 44).

3. *Titration with other reagents*

Alternative reagents to stannous and titanous salts have been used for the estimation of nitro compounds. These include vanadium salts,[1] trivalent molybdenum salts,[2] chromous chloride,[3] zinc amalgam,[4] and electrolytic cadmium in the Jones reductor.[5] The methods of estimation are similar to the above, being based on a determination of an excess of the reducing agent, or on a titration of the base. The following are typical estimations using these reagents.

(a) Nitrobenzene (0·5 g.) is sulphonated by heating with fuming sulphuric acid (20 ml.) on a water-bath for 2 hr. The solution is cautiously made up (to 250 ml.) with water; aliquot

[1] M. V. Gapchenko and O. G. Sheintsis, *Zavodskaya Lab.* 1940, **9**, 562–5; P. C. Banerjee, *J. Ind. Chem. Soc.* 1942, **19**, 35–40.
[2] M. V. Gapchenko, *Zavodskaya Lab.* 1941, **10**, 245–8.
[3] K. Someya, *Z. anorg. allgem. Chem.* 1928, **169**, 293–300.
[4] M. M. Lobunets and M. I. Per'e, *Univ. état Kiev, Bull. sci. Rec. chim.* 1936, **2**, 73–9; *ibid.* 1937, **3**, 37–41; and 43–7.
[5] M. M. Lobunets, *Zavodskaya Lab.* 1938, **7**, 872–4; *ibid.* 1937, **3**, 71–8; *ibid.* 1939, **4**, 23–36, 414.

portions (10–15 ml.) are run into a standard solution of vanadium sulphate (10–20 ml., 0·2 N) and boiled in an atmosphere of
carbon dioxide for a few minutes. The solutions are cooled and
back-titrated with ferric alum using potassium thiocyanate or
phenosafranine as the indicator. The results are more accurate
if the solution is buffered by adding Rochelle salt, or sodium
tartrate, before the back-titration.

PREPARATION: *The indicator.* A solution is made up so
that 1 drop of vanadous sulphate decolorizes 2–3 drops of the
indicator and 1 drop of ferric chloride just restores the colour.

(*b*) An excess (100–200 %) of a standard solution of a
molybdous salt is added to a flask with an atmosphere of carbon
dioxide above the nitro compound. After standing for 2 or 3 min.
the excess of the reducing agent is titrated with a standard
solution of ferric alum (see p. 187) using methylene blue
(2–3 drops) as the indicator.

(*c*) A solution of chromous chloride (an excess) is added to
the nitro compound, and after allowing the mixture to stand at
room temperature for 2–5 min. the excess is titrated with ferric
alum (N).

PREPARATION: *Chromous chloride solution.* A solution of
chromous chloride in hydrochloric acid is prepared by shaking
the trivalent chromic chloride with zinc amalgam. Ammonium
thiocyanate (10 ml., N) is added and the mixture shaken for
5 min. to remove traces of chromic ion. The amalgam is run off
and the solution is used immediately in the above estimation.
The strength of the solution depends on the reduction with zinc
amalgam, and it can be determined by titration with potassium
permanganate.

(*d*) Two methods of estimation are available for amino
compounds formed by reduction of nitro groups with zinc
amalgam.

(i) Liquid zinc amalgam (20 ml.) is added to nitrobenzene
(1 g.) in a bottle fitted with a ground-glass stopper and the
mixture is shaken until the contents are colourless. The solution
is run off, the amalgam is washed with dilute acid and the
washings added to the solution which is then made up (to

250 ml.). Aliquot portions (20 ml.) are titrated with a standard solution of potassium bromide-potassium bromate (see p. 179). The method has been used successfully for o-, m- and p-nitrobenzoic acids, m-nitrocinnamic acid, and o-, m- and p-nitrotoluenes. It is quick and easy to carry out.

Accuracy ± 0·1–0·2 %.

(ii) Aliquot portions of the solution are titrated with a standard solution of sodium nitrite using starch iodide paper as an external indicator (see p. 177).

The method has been used for the estimation of o-, m- and p-nitrotoluenes, and m- and p-nitrocinnamic acids. It cannot be used for the estimation of o-nitrocinnamic acid.

(e) A number of nitro compounds have been reduced in the Jones reductor using electrolytic cadmium and the amino compound has been estimated asd escribed in para. (d)(i) above.

B. To Ammonia

The Kjeldahl method for the determination of nitrogen in organic compounds is adopted widely on account of the ease of operation, the simplicity of the apparatus, and because it is possible to carry out a large number of estimations simultaneously. The substance is decomposed with hot, concentrated sulphuric acid and the nitrogen is fixed as ammonium sulphate. This operation is hastened by adding potassium sulphate to the mixture to raise the boiling-point. The final estimation is by distillation with excess sodium hydroxide to separate the ammonia which is collected in standard acid.[1] This method, although satisfactory for the majority of organic nitrogen compounds, has to be modified for nitro compounds owing to incomplete reduction. Several methods have been proposed using widely varying reducing agents and catalysts.[2] The

[1] J. Kjeldahl, Z. anal. chem. 1883, 22, 366–82, Compt. rend. Lab. Carlsberg, 1883, 2, [1], 12.

[2] M. Jodlbauer, Chem. Zentralblatt, 1886, [3], 17, 433; A. von Asboth, ibid. 1886, [3], 17, 161; J. W. Gunning, Z. anal. Chem. 1889, 28, 188; F. W. Dafert, ibid. 1888, 27, 224; J. Milbauer, ibid. 1902, 42, 725; M. Krüger, Ber. 1894, 27, 1633–5; C. Flamand and B. Prager, ibid. 1905, 38, 559–60; B. M. Margosches and E. Scheinost, ibid. 1925, 58, 1850–60; A. Eckert, Monatsh. 1913, 34, 1957–64; W. C. Cope, Ind. Eng. Chem. 1916, 592–3; J. Soc. chem. Ind. 1916,

reducing agents used normally are glucose, zinc dust and acid, sodium thiosulphate and salicylic acid, and less frequently sodium hydrosulphite. Selenium, mercury, mercuric oxide, copper or ferrous sulphate have been used alone or together as catalysts in the digestion, but it should be realized that it is impossible to choose a set of conditions which can be applied to all nitrogenous compounds.[1] The use of selenium has been criticized adversely owing to low values having been recorded through loss of nitrogen.[2] It is possible that no loss of nitrogen occurs if the amount of selenium does not exceed a limiting amount (0·25 g.), although under the optimum conditions low values are obtained with p-nitroaniline.[3] When mercury, or mercuric oxide, is used as the catalyst it is essential to add sodium or potassium sulphide to the reaction mixture before the final distillation with sodium hydroxide. This precaution is unnecessary when copper or ferrous sulphate are used.

The following methods have been found to give reliable results with aromatic nitro compounds of widely varying constitution.

1. *Macro-estimation.*

Sulphuric acid (30 ml., 96 %) and salicylic acid (2 g.) are added to the nitro compound (0·5 g.) in a Kjeldahl flask (approximately 500 ml.), and the solid is dissolved either by rotating the flask or by warming on a water-bath. Zinc dust (2 g., in small amounts) is added, with shaking, to the liquid which is kept near to room temperature. After the final addition the mixture is allowed to stand at room temperature for $1\frac{1}{2}$–2 hr. with shaking (at intervals of 10–15 min.) and then left overnight. The flask is clamped at an angle and heated

pp. 907–8; M. F. Lauro, *Ind. Eng. Chem.* (Anal. ed.), 1931, **3**, 401–2; C. F. Poe and M. E. Nalder, *ibid.* 1935, **7**, 189; R. A. Harte, *ibid.* 1935, **7**, 432–3; A. Elek and H. Sobotka, *J. Amer. Chem. Soc.* 1926, **48**, 501–3; Z. Csuros, E. Fodor-Kencsler and I. Gretists, *Magyar-Chem. Folyoirat*, 1942, **48**, 33–42; cf. *Chem. Zentralblatt*, 1943, **1**, 545.

[1] P. L. Kirk, *Anal. chem.* 1950, **22**, 354–8.

[2] R. M. Standstedt, *Cereal Chem.* 1932, **9**, 156; C. F. Davis and M. Wise, *ibid.* 1933, **10**, 488–93; S. R. Snider and D. A. Coleman, *ibid.* 1934, **11**, 414–30; R. A. Osborn and A. Krasnitz, *J. Assoc. Official Agr. Chem.* 1934, **17**, 339–42; C. O. Willits, M. R. Coe and C. L. Ogg, *ibid.* 1949, **32**, 118–27; C. L. Ogg and C. O. Willits, *ibid.* 1950, **33**, 100–3.

[3] R. B. Bradstreet, *Ind. Eng. Chem.* (Anal. ed.), 1940, **12**, 657.

gently in a fume cupboard until no more gases are evolved ($1\frac{1}{2}$–2 hr.) and then more strongly during $1\frac{1}{2}$–2 hr. so that the reaction mixture boils. The digestion is completed by cooling slightly, adding yellow mercuric oxide (1 g.), boiling for $1\frac{1}{2}$–2 hr., cooling, adding potassium sulphate (7·5 g.) and concentrated sulphuric acid (10 ml.) and boiling for a further $1\frac{1}{2}$–2 hr. If the resulting solution is clear and colourless the estimation is continued, otherwise more potassium sulphate (1 g.) is added and the heating continued for $\frac{1}{2}$–1 hr. After allowing the reaction mixture to cool, water (200 ml., distilled) and a solution of potassium sulphide (25 ml., 80 g. in 1 l. of distilled water) are added to dissolve the solid. The solution is transferred to a distillation apparatus, granulated zinc (1 g.) is added to give a steady stream of hydrogen during the distillation and so prevent 'bumping'. The apparatus is assembled and a solution of sodium hydroxide (85–90 ml., 750 g. in 1 l. of distilled water) is run into the solution immediately before the distillation is started. The determination is completed in the normal manner.

Numerous types of apparatus have been proposed for the final distillation, but whichever form is used it is essential it should include a splash bulb to prevent alkali being carried over into the standard acid.

2. *Micro-estimation.*

Concentrated sulphuric acid (4 ml.) is added to a mixture of the nitro compound (containing about 2·5 mg. of nitrogen), glucose (20 mg.), copper sulphate (20 mg.), small crystals of potassium sulphate (1–1·5 g.), and a few pieces of well-washed alundum. The mixture is digested over a micro-burner in an almost horizontal position. When a homogeneous liquid is obtained (20–30 min.) selenium oxychloride (1 drop) is added and heating continued for a further 15 min. The distillation is done as described below or as given in *Quantitative Organic Micro-analysis based on the methods of Fritz Pregl.*[1]

[1] A. C. Andersen and B. N. Jensen, *Z. anal. Chem.* 1931, **83**, 114–20; J. Grant, *Quantitative Organic Micro-analysis based on the methods of Fritz Pregl*, 5th English ed., 1951, pp. 95–104.

APPARATUS. The apparatus[1] (fig. 26) is similar to that[2] described by Pregl[3] with changes in structural detail which give it a compact and robust form.

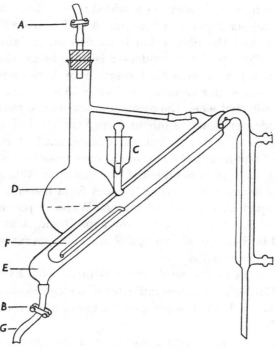

Fig. 26.

The screw clips A and B are opened and steam is blown through the apparatus for half an hour before it is used. This can be done conveniently during the digestion of the sample with sulphuric acid. The digest is cooled and transferred quantitatively, through the funnel C, into the reaction vessel F, the Kjeldahl flask is washed with water (4 lots of 2·5 ml. each), the washings are added to F and an aqueous solution of sodium hydroxide (12 ml., 50 %) is poured into C. Water is boiled in

[1] R. Markham, *Biochem. J.* 1942, **36**, 790–1.
[2] K. I. Parnas and R. Wagner, *Biochem. Z.* 1921, **125**, 253–6.
[3] J. Grant, *Quantitative Organic Micro-analysis based on the methods of Fritz Pregl*, 5th English ed., 1951, pp. 95–104.

the flask D and, by closing the screw clip A, steam is forced into the outer jacket E where it condenses and flows from the apparatus through G until the liquid in the reaction vessel is at 100° C. when steam passes out. The screw clip B is closed and the sodium hydroxide solution is added to the digest by raising the glass stopper slightly in the cup. A seal is made by adding water to C and the distillation is carried out at about 5 ml. a minute. The tip of the condenser is kept below the surface (about 1 mm.) of a saturated solution of boric acid (about 10 ml., see below) in a receiver until some 15 ml. of the distillate has been collected when the receiver is lowered so that the end of the condenser is above (about 3 cm.) the level of the liquid in the receiver. The distillation is continued until a further 5 ml. is collected and finally the end of the condenser is washed with water and the washings added to the distillate. When the distillation is finished the burner is taken from beneath the flask and the liquid in F is sucked into E. Water is poured into C and washes through F into E. The clips B and A are opened, the liquid in E is run off through G and the apparatus is ready for a second estimation.

The contents of the receiver are titrated with hydrochloric acid (0·025 N) using a mixed indicator of methyl red and methylene blue (see below). The end-point is taken when the indicator turns violet.

The ammonia is usually absorbed in a saturated solution of boric acid although it is reported that absorption in nickel ammonia sulphate gives a sharper end-point in the final titration with acid.[1] Care is necessary to prevent loss of ammonia during absorption and it is claimed that losses are reduced[2] if a Goessman trap[3] is used as the receiver.

The apparatus may be used for a sample containing as little as 0·02 mg. of nitrogen when 10 ml. only of the distillate is collected and the tip of the condenser is not placed below the surface of the liquid in the receiver during distillation.

[1] J. Blom and B. Schwarz, *Acta Chem. Scand.* 1949, **3**, 1439–40.
[2] R. M. Silverstein and R. Perthell, *Anal. Chem.* 1950, **22**, 949–50.
[3] C. I. Goessman, *Massachusetts Agric. Exptl. Sta.* 1898, Bulletin 54.

For the determination of nitrogen in nitro azo dyes pure, bleached cotton can be substituted for glucose.[1]

PREPARATIONS: *Boric acid.* A saturated solution (approx. 4 %) is made by shaking boric acid and distilled water in a bottle.

The indicator. Methyl red (0·125 g.) and methylene blue (0·083 g.) are dissolved in absolute alcohol (100 ml.). When the indicator is to be used over a long period it is better to prepare separate solutions of methyl red and methylene blue in alcohol and make up the mixed indicator each week from the two solutions.

COLORIMETRIC
A. Specific for Mononitroparaffins

The *aci* form of mononitroparaffins can be estimated colorimetrically with ferric chloride, and if the reagent is added immediately after the complete formation of the *aci* form the amount of the nitroparaffin is determined.[2] The colour is transient, and accurate determinations are possible only if the unknown and standard solutions are prepared simultaneously under identical conditions. The optimum pH for the production of maximum colour is 1·25–1·30 (see below). Interference from the yellow colour of ferric chloride can be minimized by reducing the volume added when only a small amount of the nitroparaffin is present. The method has been used successfully for the estimation of nitroethane, nitropropane and nitrobutane.

Aqueous sodium hydroxide (1·5 ml., 20 %) is added to an alcoholic solution of the nitro compound (1–15 ml., containing 1–20 mg.) in a graduated flask (25 ml.) and allowed to stand at room temperature for 15 min. The mixture is acidified (6 ml., 1 part concentrated hydrochloric acid, 7 parts water) and a solution of ferric chloride (0·5 ml., 10 %) is added immediately. After leaving to stand for 15 min. the solution is made up to the

[1] R. V. Bhat, *Proc. Indian Acad. Sci.* 1941, **13** A, 269–72.; cf. R. B. Forster, P. R. Mehta and K. Venkataraman, *J. Soc. Dyers Col.* 1938, **54**, 210.

[2] E. W. Scott and J. F. Treon, *Ind. Eng. Chem.* (Anal. ed.), 1940, **12**, 189–90.

mark, and after mixing compared in a colorimeter with a standard solution.

Mononitromethane can be estimated colorimetrically using vanillin and ammonia.[1]

B. Specific for Aromatic Nitro Compounds

A small amount of an aromatic nitro compound can be estimated colorimetrically by the addition of a dilute solution of sodium hydroxide and comparison with standard solutions prepared under identical conditions.[2] The unknown and standard solutions should be made up simultaneously as the colour fades rapidly. The method has been found to be satisfactory for the analysis of aromatic nitro compounds in air; trinitrotoluene has been estimated in concentrations as low as 0·5 mg. in 100 l. of air.

Air (50–100 l.) is drawn through three wash-bottles filled with methyl or ethyl alcohol. The contents of the first two bottles are mixed and an aqueous solution of sodium hydroxide (2 ml., 2 %) is added. The colour is compared with standard solutions after 10 min.

MISCELLANEOUS

(1) Nitro compounds can be estimated in a polarograph.[3] A solution of the nitro compound (0·0005–0·0167 M) is electrolysed in sulphuric acid (0·05 M). The reduction potential probably corresponds to the formation of the hydroxylamine, as only one stage is observed. The experimental conditions for the estimation must be controlled very carefully as a change of 1 mm. in pressure on the dropping mercury cathode introduces an error of 3 %, and a variation of 2° C. in the temperature causes the current to fluctuate by 3 %.

[1] C. D. Manzoff, *Z. nachr.- u. Genussm.* 1914, **27**, 469; W. F. Machle, E. W. Scott and J. F. Treon, *loc. cit.* (p. 195, n. 2).

[2] E. Elvove, *J. Ind. Eng. Chem.* 1919, **11**, 860–4; cf. C. A. Lobry de Bruyn, *Rec. trav. chim.* 1895, **14**, 89–94 and 151–5; A. Hantzsch and H. Kissel, *Ber.* 1899, **32**, 3137–48; J. Meisenheimer and K. Witte, *ibid.* 1903, **36**, 4164–74; J. Meisenheimer, *Ann.* 1902, **323**, 205–46.

[3] T. de Vries and R. W. Ivett, *Ind. Eng. Chem.* (Anal. ed.), 1941, **13**, 339–40; J. Haslam and L. H. Cron, *J. Soc. Chem. Ind.* 1944, **63**, 94–5T.

(2) Dinitrobenzene, dinitrochlorobenzene, dinitronaphthalene, dinitrotoluene and dinitroxylene have been estimated colorimetrically in the presence of the corresponding mononitro compounds by the colour produced when a dilute alcoholic solution of sodium hydroxide (freshly prepared from sodium oxide) is added to a solution in acetone.[1]

(3) A gravimetric estimation has been described in which the nitro group is reduced to the amino group with tin and hydrochloric acid so that atmospheric oxidation is prevented, and the evolution of gaseous hydrogen is minimized by controlling the concentration of the acid. Under these conditions the weight of tin used is proportional to the amount of nitro group present. After reaction the excess of tin is weighed. Iodo- and aldehydic groups interfere.[2]

CYANO COMPOUNDS

The well-known reactions used for the estimation of inorganic cyanides cannot be adapted to the quantitative analysis of the cyano group, as this latter class of compound does not give the cyanide ion in solution. Two methods, other than the determination of the nitrogen content by combustion, have been described and are as follows:

Hydrolysis to the corresponding amide:
$$R.CN + H_2O = R.CONH_2.$$
Formation and estimation of ammonia.

HYDROLYSIS TO THE CORRESPONDING AMIDE
$$R.CN + H_2O = R.CONH_2.$$

The quantitative hydrolysis of a cyanide to an amide can be used for estimation.[3] Water and a catalyst are added to the cyanide, and after the hydrolysis is completed, the water remaining is determined by titration with the Karl Fischer

[1] F. L. English, *Anal. Chem.* 1948, **20**, 745–6.
[2] C. E. Vanderzee and W. F. Edgell, *ibid.* 1950, **22**, 572–4.
[3] J. Mitchell, Jr. and W. Hawkins, *J. Amer. Chem. Soc.* 1945, **67**, 777–8.

reagent (see pp. 50–5). As in other estimations with this reagent a blank estimation is made, and the difference between the two readings is equivalent to the cyanide present.

Dilute sulphuric acid and alkalis have been used unsuccessfully for the hydrolysis. The former gives reproducible results but the accuracy is poor, and alkalis react with the Karl Fischer reagent and interfere seriously with the estimation at the concentrations necessary for complete hydrolysis. Boron trifluoride in acetic acid and water has none of these disadvantages, and accurate results are obtained generally with the normal lower aliphatic cyanides, the cyanides of dibasic acids and many of the aromatic compounds. Any configurational peculiarity, such as the 'ortho effect', can prevent the quantitative formation of the amide. o-Tolunitrile and α-naphthonitrile give low results, although m- and p-tolunitrile and β-naphthonitrile react quantitatively. Methyleneacetonitrile and cyanoacetic acid also give low values.

Little interference is found with amides, but with alcohols the hydroxyl group is esterified, quantitatively eliminating an equivalent of water. This may be corrected by determining the hydroxyl content independently (see pp. 41–72).

The hydrolysing agent (20 ml.) is added to the cyanide (up to 10 m.equiv.) and weighed into a glass-stoppered flask (250 ml.). Two blanks are prepared, and the flasks are sealed tightly (if possible the stoppers should be held by spring clips) and placed in a water-bath at $80 \pm 2°$ C. for 2 hr. The flasks are taken from the water-bath, cooled to room temperature and then placed in a bath of finely chopped ice during the addition of dry pyridine (15 ml.). The homogeneous mixture is titrated finally with the Karl Fischer reagent.

As the cyanide grouping does not interfere the original sample should be titrated with the Karl Fischer reagent to determine the amount of water present.

Accuracy $\pm 0.3 \%$.

PREPARATION: *The Karl Fischer reagent* (see p. 55).

FORMATION AND ESTIMATION OF AMMONIA

The standard Kjeldahl procedure for the estimation of nitrogen is not used for the accurate analysis of cyanides, but a preliminary digestion with hydriodic acid is done and the modified procedure of Kjeldahl-Gunning-Arnold (cf. pp. 191–4) is adopted. A mixture of potassium iodide and sulphuric acid is usually preferred to the free acid owing to the rapid deterioration of the latter once the container is opened. It has been reported recently that the preliminary reduction is unnecessary and accurate values are obtained using the ordinary Kjeldahl procedure.[1]

A limited number of cyanides, including saturated and unsaturated aliphatic, aromatic and substituted (bromo- and nitro-) aromatic compounds have been analysed successfully after reduction.[2] Two of these, acrylonitrile and acetonitrile, have boiling-points below 100° C., but both can be analysed by the standard procedure without any loss of accuracy. The results normally compare well with those found by Friedrich's modification of the Kjeldahl method which is frequently taken as a standard.[3] When both nitro- and cyano-groups are present the total nitrogen is estimated, but all groups containing nitrogen are not converted quantitatively. Azo compounds have been examined and sometimes only part of the nitrogen is recovered The method is unsatisfactory when a large amount of water is present, although a few millilitres has no effect.

A mixture of the cyanide (containing 40–60 mg. of nitrogen), potassium iodide (1·5 g.) and concentrated sulphuric acid (30 ml.) is heated with occasional shaking in a Kjeldahl flask on a water-bath for 45 min. Potassium sulphate (10 g.), anhydrous copper sulphate (0·3 g.), selenium (0·1 g.) and a few pieces of unglazed porcelain are added and the heating continued, gently at first and then boiling vigorously over a free flame, for 1 hr. after the mixture has become clear green in colour. The volume must not be allowed to fall below 20 ml., and any iodine

[1] C. H. Vanetten and B. M. Wiele, *Anal. Chem.* 1951, **23**, 1338–9.

[2] E. L. Rose and H. Ziliotto, *Ind. Eng. Chem.* (Anal. ed.), 1945, **17**, 211–2.

[3] A. Friedrich, E. Kühaas and R. Schnürch, *Z. physiol. Chem.* 1933, **216**, 68–76.

remaining in the neck of the flask at the end of this time is removed by gentle heating with a flame. The mixture is cooled, water (250 ml.) is added, and the distillation done in the normal manner (see pp. 192–5). The solution is made strongly alkaline by the addition of sodium hydroxide solution (90 ml. is usually enough, 50 %).

Percentage of nitrogen

$$= \frac{(\text{ml. of } H_2SO_4 \times N - \text{ml. of NaOH} \times N) \times 1 \cdot 401}{\text{weight of sample}}.$$

Accuracy $\pm 0 \cdot 5 \%$.

NITROSO COMPOUNDS

The quantitative reduction methods of analysis which are used for the nitro grouping can also be used without change for nitroso compounds. Reduction to the corresponding base by reagents such as stannous chloride or titanous chloride, or to ammonia as in Kjeldahl's method, therefore can be used.

Reduction:

A. To the corresponding primary base:

$$R.NO + 4H = R.NH_2 + H_2O.$$

B. To ammonia.

Miscellaneous.

REDUCTION

Nitroso compounds can be estimated by reduction either to the corresponding primary base or to ammonia as described under nitro compounds (see pp. 184–95).

MISCELLANEOUS

Nitroso compounds liberate an equivalent of iodine from an acidified solution of potassium iodide, and estimation is possible therefore by titration of the reaction mixture with a standard solution of sodium thiosulphate.[1]

[1] M. M. Lobunets and E. N. Gortins'ka, *Univ. état Kiev Bull. Sci. Rec. chim.* 1939, **4**, 37–9.

Nitrosobenzene (1 g.) is dissolved in ethyl alcohol (150 ml.), and the solution is diluted (to 250 ml.) with water. Measured volumes (25 ml.) are added to hydrochloric acid (15–20 ml., 6N) and potassium iodide (10 ml., 20%), and after a few minutes the iodine is titrated with a standard solution of sodium thiosulphate (0·1N), using starch (1 ml., 1%) as an indicator near the end-point.

Accuracy $\pm 0.5\%$.

ISOCYANATES AND ISOTHIOCYANATES

A smooth and quantitative reaction takes place between an amine, or an amino compound, and either an isocyanate or isothiocyanate. A method of estimation based on this reaction has been used for a limited number of isocyanates and isothiocyanates and should be useful as a general method for these classes of compound:

$$R.N{=}C{=}O + C_4H_9NH_2 = R.NH.CO.NH.C_4H_9,$$
$$R.N{=}C{=}S + C_4H_9NH_2 = R.NH.CS.NH.C_4H_9.$$

An excess of an amine, or an amino compound, is added to the isocyanate or isothiocyanate, and after complete reaction the mixture is titrated with a standard solution of acid.[1] n-Butylamine has been found to be a convenient reagent as it is reasonably reactive, and yet its boiling-point (78° C.) is high enough to stop losses from evaporation. A non-hydroxylic solvent must be used, and dioxane therefore is very suitable as it fulfils this condition, and as it is also miscible with water the final titration with a standard aqueous solution of acid is straightforward. Impurities with an active hydrogen, basic or acidic substances and thiocyanates which isomerize to isothiocyanates on heating interfere with the estimation. The acidic or basic constituents may be found separately by titration and allowed for in the estimation.

The isocyanate, or isothiocyanate (0·002 mol.), is added to a solution of n-butylamine (20 ml., see below) in a glass-stoppered Erlenmeyer flask (100–150 ml.). The contents are

[1] S. Siggia and J. G. Hanna, *Anal. Chem.* 1948, **20**, 1084.

mixed by swirling the flask and left to stand at room temperature (for 15–45 min.). Aliphatic isocyanates and isothiocyanates react slower than the aromatic compounds, and some may have to be heated under a reflux condenser for a few minutes to finish off the reaction. When this is necessary the reflux condenser is washed with water and the washings added to the flask. The excess of n-butylamine is found by titration with sulphuric acid (0·1 N) using methyl red as the indicator.

The strength of the reagent (20 ml.) is found by titration with sulphuric acid (0·1 N) using methyl red as the indicator.

Percentage of isocyanate

$$= \frac{\text{ml. of acid (blank} - \text{sample)} \times \text{mol. wt. of isocyanate} \times 100}{\text{weight of isocyanate} \times 1000}.$$

Accuracy $\pm 0·5 \%$.

PREPARATION: *n-Butylamine solution.* Dioxane is stirred over pellets of sodium hydroxide which are changed each day until the pellets no longer turn brown. The dioxane is then distilled from and kept over solid sodium hydroxide. The solution of n-butylamine is made by diluting the base (12·5 g.) with dioxane (to 100 ml.).

CHAPTER VII

OTHER GROUPS

ACETYL AND BENZOYL GROUPS

ESTIMATION BY HYDROLYSIS

$$R.O.COCH_3 + H_2O \longrightarrow R.OH + CH_3.COOH$$

$$\begin{matrix} R \\ \diagdown \\ R' \end{matrix} N.COCH_3 + H_2O \longrightarrow \begin{matrix} R \\ \diagdown \\ R' \end{matrix} NH + CH_3COOH.$$

The simple determination of acetyl and benzoyl groups attached to oxygen, as in esters, or to nitrogen, as in anilides, is possible either by acid or alkaline hydrolysis followed by distillation and titration of the acetic acid or benzoic acid which is formed.[1] If the acetyl or benzoyl group is attached to oxygen, hydrolysis is possible by heating for $\frac{1}{2}$ hr. either with sulphuric acid or p-toluenesulphonic acid or for 5 min. with alcoholic sodium hydroxide. There are exceptions which take longer heating, sometimes up to $2\frac{1}{2}$ hr. with alkali in alcohol as with the methyl ester of triacetylcholic acid. When groupings are combined with nitrogen the reaction is carried out usually with alcoholic sodium hydroxide. The ease of hydrolysis of an acetyl or benzoyl group depends upon the structure of the compound and also on the solubility in the hydrolysing agent. When the material is insoluble in an aqueous or alcoholic solution of acid or alkali it is dissolved in pure pyridine (1 ml.), and sodium hydroxide in methyl alcohol is used as the saponifying agent. When hydrolysis is completed the acetic or benzoic acid is separated from the acid reaction mixture by distillation through

[1] R. Kuhn and H. Roth, *Ber.* 1933, **66**, 1274–8; A. Friedrich and S. Rapoport, *Biochem. Z.* 1932, **251**, 432–46; E. P. Clarke, *Ind. Eng. Chem.* (Anal. ed.), 1936, **8**, 487–8 and 1937, **9**, 539; A. Elek and R. A. Harte, *ibid.* 1936, **8**, 267–70; R. B. Bradbury, *Anal. Chem.* 1949, **21**, 1139–42; K. Freundenberg and M. Harder, *Ann.* 1923, **433**, 230–3; K. Freundenberg and E. Weber, *Z. angew. Chem.* 1925, **38**, 280–5; F. Pregl and A. Soltys, *Mikrochemie*, 1929, **7**, 1–9; A. G. Perkin, *J. Chem. Soc.* 1905, **87**, 107–10; cf. F. Kögl and J. J. Postowsky, *Ann.* 1924, **440**, 34–5; A. J. Bailey and R. J. Robinson, *Mikrochemie*, 1934, **15**, 233–6.

a transparent quartz condenser and estimated by titration with alkali or iodometrically. Phosphoric acid must not be used for the hydrolysis, unless a special form of apparatus is used, as it is always found in the distillate.[1] Sulphuric acid may give rise to sulphur dioxide, but both sulphur dioxide and carbon dioxide can be expelled completely from the solution without losing acetic acid by boiling with sulphuric acid for 7–8 sec. A cold finger can be placed in the neck of the Erlenmeyer flask to ensure that no acetic acid is lost.[2] An alternative procedure is by acid saponification in ethyl alcohol.[3] The ethyl acetate is separated by distillation and then hydrolysed with an excess of alkali which is back-titrated. In the original macro-procedure, errors were possible through loss of volatile ester and by the formation of acetic acid by oxidation of ethyl alcohol. The method has been adapted to the semi-micro- and to the micro-scale.

The saponifying agent (see below) is chosen after the solubility of the acetyl or benzoyl compound in the four saponifying agents is found roughly by testing with small samples in test-tubes. The substance (to neutralize about 5 ml. of 0·01 N alkali) is weighed into the reaction flask (fig. 27) using a long-handled weighing tube or small boat for solids and a capillary tube for liquids. Solids which dissolve only with difficulty are powdered as finely as possible in an agate mortar before weighing. The hydrolysing agent (1 ml. of sulphuric acid, or 1 ml. of p-toluenesulphonic acid, or 1 ml. of 5 N sodium hydroxide, or 4 ml. of sodium hydroxide in methyl alcohol, see below) is added from a pipette, the funnel at F is closed, the joint is sealed with water (1–2 ml.) and the apparatus is assembled as shown. Liquids are weighed into the flask containing the reagent and with the condenser in position. The capillary tube is placed point downwards in the reagent and broken by pressing with a glass rod. The rod is washed with water (1 ml.) or with the hydrolysing agent (1 ml.). The condenser is fixed in a vertical

[1] E. Weisenberger, *Mikrochemie ver. Mikrochim. Acta*, 1947, **33**, 51–69.
[2] G. Kainz, *ibid.* 1950, **35**, 400–6.
[3] A. G. Perkin, *loc. cit.* (p. 203, n. 1); E. Weisenberger, *Mikrochemie ver. Mikrochim. Acta*, 1942, **30**, 241–59.

position, the bubble counter is connected with an oxygen or nitrogen supply and the rate of flow is adjusted (to about 50 bubbles per min.). The reaction flask is placed in a bath of boiling water so that the side limbs B and D are covered. The reaction mixture is kept at its boiling point for half an hour, the water bath is then removed, the flask is left to cool to room temperature and the seal at F is opened. The reflux condenser is rinsed into the flask with water (about 5 ml.), removed and washed thoroughly with water and replaced as a distilling

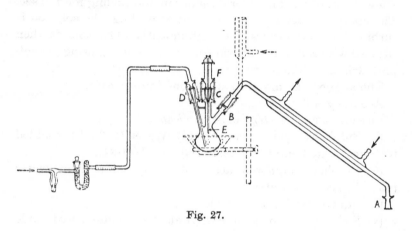

Fig. 27.

condenser. If methyl alcohol is used in the hydrolysing agent or if pyridine is used as a solvent it is removed by distillation (the first 5 ml. are collected and rejected). The reaction mixture is prepared for the final distillation by adding, through the funnel F, (a) sodium hydroxide (1·0 ml., 5N), after hydrolysis by sulphuric acid, (b) sodium hydroxide (0·5 ml., 5N), after hydrolysis by p-toluene sulphonic acid, (c) sulphuric acid (1·0 ml., Wenzel's acid, see below) after hydrolysis by alkali. One or two pieces of pumice are added to the flask, the funnel F and the glass rod are washed with water (about 5–7 ml.). The bubble counter is connected and the acetic or benzoic acid is distilled out of the mixture (about 1 ml. is collected per minute), using a small flame and a Babo funnel or asbestos gauze. The distillate is collected in a measuring cylinder (25 ml.), and in order to drive the acid over completely the liquid in the flask is reduced to a small volume (2–4 ml., about 5 ml. of distillate) before more water (5 ml.) is added slowly from the funnel F. The distillation is not interrupted and the procedure is repeated (until 20 ml. are collected). The contents of the cylinder are poured into a clean flask. The distillation is continued and further fractions (15 ml. each) are collected until no more acid distils over (usually 4 fractions, 30–40 min.).

Each fraction is tested for the absence of sulphate by adding a few drops of barium chloride solution and boiling for 7–8 sec. The analysis must be rejected if, after cooling, the solution is turbid, but this occurs very infrequently. The acid is then titrated with sodium hydroxide or baryta (0·01 N), using phenolphthalein as the indicator.

The analysis should always be done in duplicate.

Accuracy $\pm\,0\cdot3$–$0\cdot5\,\%$.

PREPARATIONS: *Hydrolysing agents.*

(a) Sulphuric acid (Wenzel's acid). Water (200 ml.) is added to concentrated sulphuric acid (100 ml., $d = 1\cdot84$).

(b) p-Toluenesulphonic acid. A 25 % solution in water (43 g. in 100 g. of water).

(c) Aqueous sodium hydroxide, 5 N.

(d) Sodium hydroxide in methyl alcohol. Sodium hydroxide (4 g.) is dissolved in aqueous methyl alcohol (100 ml., 50:50). The methyl alcohol is freed from traces of acid by refluxing for 15 min. with solid potassium hydroxide pellets and then distilling.

Apparatus.[1] The three necks of the reaction flask (capacity 50–60 ml.) are fitted with B 10 joints. The gas inlet tube (2 mm. bore) is connected to the limb D (80 mm. long) through a bubble-counter, filled with potassium hydroxide (50 %) and a U-tube filled with soda-lime granules. The central neck C (80 mm. long) holds a funnel F (8 ml. capacity when the rod E is in position). The third limb B (65 mm. long, 50 mm. bore) is at an angle of 50° to the centre limb and supports a quartz condenser (centre tube 35–40 cm. long). The condenser is bent about 5 cm. above

[1] R. Kuhn and H. Roth, *loc. cit.* (p. 203, n. 1); cf. E. Weisenberger, *loc. cit.* (p. 204, n. 1).

the joint A at 40°, so that by connecting A to B it may be used in the reflux position. All the joints are moistened with water and are held securely by springs during the estimation.

MISCELLANEOUS

A simple method of estimation for the acetyl groups in glucosides and acetates of sugars is to shake them with ethyl alcohol (a few ml.) and a known volume of aqueous sodium hydroxide (4 ml., 0·045N) for 4–24 hr. The excess alkali is titrated with sulphuric acid (0·05N), using phenolphthalein as the indicator.

A modification of this method has been used for O-acyl derivatives.[1] The solid is dissolved in acetone and hydrolysed at room temperature by shaking with an aqueous solution of sodium hydroxide (0·01N). The excess alkali is titrated with sulphuric acid (0·01N) using phenol red as the indicator.

METHOXYL AND ETHOXYL GROUPS

ESTIMATION BY HYDRIODIC ACID

$$R.OCH_3 + HI \longrightarrow CH_3I + R.OH,$$
$$R.OC_2H_5 + HI \longrightarrow C_2H_5I + R.OH.$$

Methoxyl and ethoxyl groups are determined by heating the substance with constant boiling hydriodic acid and estimating the methyl or ethyl iodide which is formed.[2] The volatile alkyl halide is removed from the boiling reaction mixture in a slow stream of carbon dioxide, washed free from hydrogen iodide, iodine and hydrogen sulphide and is then estimated either gravimetrically or volumetrically. In the former the halide is collected in an alcoholic solution of silver nitrate to precipitate the double salt, silver iodide-silver nitrate, which is filtered off, decomposed with dilute nitric acid into silver nitrate which goes into solution and insoluble silver iodide which is removed and weighed.

Two volumetric procedures have been described. In one the alkyl halide is absorbed in pyridine and the corresponding

[1] J. F. Alicino, *Anal. Chem.*, 1948, **20**, 590.
[2] S. Zeisel, *Monatsh.* 1885, **6**, 989–96; 1886, **7**, 406–9.

pyridinium iodide is titrated either with mercuric oxycyanide[1] or with silver nitrate.[2] In the second the iodide is taken up in a solution of bromine, sodium acetate and acetic acid when the following reactions take place:

$$R.I + Br_2 \longrightarrow R.Br + IBr, \quad IBr + 3H_2O + 2Br_2 \longrightarrow HIO_3 + 5HBr.$$

The sodium acetate neutralizes the hydrogen bromide, and the equivalent of iodate is estimated by adding potassium iodide and titrating the iodine with sodium thiosulphate.[3]

The volumetric procedure is preferable to the gravimetric estimation for almost all classes of compounds, as it is quicker and accurate when reagents are used which give good blank tests (use up not more than 0·2 ml. of the sodium thiosulphate solution). If the percentage of alkoxyl is very low, or when the estimation is for traces of alcohol of crystallization, the gravimetric method should be used, as it is easier and more accurate to weigh a small amount of silver iodide than form and titrate the corresponding quantity of iodate. The gravimetric method fails for substances containing sulphur, as it is impossible to wash hydrogen sulphide completely from the gases passing into the receiver. The volumetric procedure can be used for these compounds.

The method as described below is useful only for methoxyl and ethoxyl groups, for although the higher alkyl iodides are formed in the same way they are not volatile enough to be estimated in the same apparatus (see p. 216). It is necessary sometimes to distinguish methyl from ethyl iodide, and this can be done by passing the gases from the reaction into an alcoholic solution of trimethylamine. A quaternary ammonium salt is formed, and if it is tetramethylammonium iodide a precipitate forms, whereas the alternative product of trimethylethylammonium iodide is soluble under the same conditions.[4]

[1] G. Ingram, *Analyst*, 1944, **69**, 265–9.
[2] A. Kirpal and T. Bühn, *Monatsh.* 1915, **36**, 853–63; K. Bürger and F. Baláž, *Angew. Chem.* 1941, **54**, 58–9.
[3] F. Vieböck and C. Brecher, *Ber.* 1930, **63**, 3207–10; T. White, *Analyst*, 1943, **68**, 366–8.
[4] W. Küster and W. Maag, *Z. physiol. Chem.* 1923, **127**, 190–5; R. Willstätter and M. Utzinger, *Ann.* 1911, **382**, 148–50; M. Phillips and M. J. Goss, *J. Assoc. Official Agr. Chem.* 1937, **20**, 292–7; L. M. Cooke and H. Hibbert, *Ind. Eng. Chem.* (Anal. ed.), 1943, **15**, 24–5.

Inconsistent or low results may arise from only partial decomposition of the substance when heated with hydriodic acid. This is due frequently to incomplete dissolution and can be overcome by adding a small amount of phenol and either propionic or acetic anhydride to the reaction mixture. It has been suggested that complete solution is made more certain by

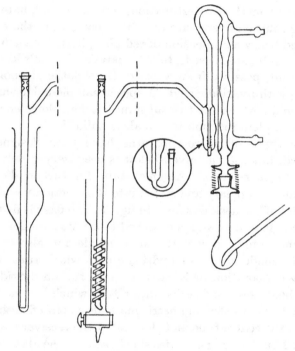

Fig. 28.

sealing the sample holder with a thin film of tartaric acid which makes it possible to bring the substance below the surface of the mixture in the flask before any reaction can take place.[1] Care must be taken to choose a suitable washing solution as some of the alkyl halide may dissolve. Methyl iodide is sufficiently soluble in sodium thiosulphate to make results appreciably low, but the effect can be minimized either by adding cadmium

[1] D. O. Hoffman and M. L. Wolfrom, *Anal. Chem.* 1947, **19**, 225–8.

sulphate to the solution or by dissolving the sodium thiosulphate in a saturated solution of sodium chloride.[1] Ethyl iodide, however, is practically insoluble in a solution of sodium thiosulphate. Alternative washers which are used include a suspension of red phosphorus in water or in dilute sodium carbonate (0·5 %), but they do not remove completely traces of hydrogen sulphide which is the cause sometimes of high results in the gravimetric method of estimation. Cadmium sulphate in the washing solution overcomes this difficulty, and in the methods described below a suspension of red phosphorus in a solution of cadmium sulphate is used. In the analysis of acetals low values often result, possibly from a loss of the volatile alcohol before reaction with hydriodic acid. A special sample holder has been described in which all the substance is held long enough to ensure complete reaction to the alkyl halide.[2]

Many types of micro-apparatus have been designed and described, but the one given below is used very widely and is known to be reliable.[3] An inlet tube for carbon dioxide is drawn into a capillary and sealed into the reaction vessel at the lower end of the neck as shown in fig. 28. The flask is fitted with a ground-glass joint and is attached to a reflux condenser with three small bulbs to cool the alkyl halide and also to ensure complete condensation of hydriodic acid which rises into the tube by evaporation or is carried up by carbon dioxide.[4] The tube is bent down and sealed into a trap which is joined to the delivery tube. In the volumetric methods a spiral fills the space between the centre tube and the wall of the receiver and forces the small bubbles of gas to rise slowly so that the alkyl halide is absorbed quantitatively in the solution. In the gravimetric method it is easier to remove the solid completely if a straight

---

[1] E. P. White, *Ind. Eng. Chem.* (Anal. ed.), 1944, **16**, 207–8; cf. A. Friedrich, *Mikrochemie*, 1929, **7**, 195–201.

[2] D. O. Hoffman and M. L. Wolfram, *loc. cit.* (p. 209, n. 1).

[3] A. Elek, *Ind. Eng. Chem.* (Anal. ed.), 1939, **11**, 174–7; J. Grant, *Quantitative Organic Micro-analysis based on the methods of Fritz Pregl*, 5th English ed., pp. 182–94; D. R. Rigakos, *J. Amer. Chem. Soc.* 1931, **53**, 3903–4; E. P. Clark, *ibid.* 1929, **51**, 1479–83; F. Neumann, *Ber.* 1937, **70**, 734–6; J. J. Chinoy, *Analyst*, 1936, **61**, 602–3; A. F. Colson, *ibid.* 1933, **58**, 594–600; M. Lieff, E. Marks and G. F. Wright, *Canad. J. Res.* 1937, **15**, 529–31.

[4] Cf. E. V. White and G. F. Wright, *ibid.* 1936, **14**, 427–9.

delivery tube and narrow receiver is used (see below). Two constrictions are made in the head of the delivery tube, and before each estimation a small drop of water is placed on the upper one and the tube is closed with a cork to form an effective seal against halide vapours. The apparatus is made from Pyrex-type tubing of 5 mm. bore.

The following modifications have been described recently. Bumping or surging of the reaction mixture can be stopped by passing the carbon dioxide through a capillary heated at 200° C. before it enters the flask. The hot gas keeps the reaction mixture boiling smoothly.[1] An apparatus which can be left unattended has been described and is useful for multiple determinations.[2] A procedure involving a double distillation in an all-glass apparatus is described for compounds which break down immediately on contact with hydriodic acid and for substances which usually give low results.[3]

It is only possible to get a reliable result by starting with a perfectly clean and dry apparatus. This is done by standing the apparatus in a chromic acid-sulphuric acid bath for at least 5 min. and then successively drawing through tap water, distilled water, acetone and air by attaching a water pump to the side tube and closing the filling tube of the trap. The apparatus is wiped on the outside and dried at 120° C. It is unnecessary to clean the whole apparatus in chromic acid before each analysis, but whenever the gravimetric method is used the delivery tube and receiver should be treated in this way to remove all fat. If this is not done it may be difficult to transfer the silver iodide quantitatively.

It is essential that the substance should dissolve completely in the reaction mixture, and so a preliminary test-tube experiment is done to find the approximate solubility. Two or three crystals of the solid are added to the same amount of phenol and one or two drops of propionic anhydride, and the mixture warmed in a water-bath at 50–60° C. If the compound dissolves no further experiment is needed, but if not the mixture is

[1] P. Saccardi, *Ann. chim. applicata*, 1944, **34**, 18–19.
[2] A. A. Houghton and H. A. B. Wilson, *Analyst*, 1944, **69**, 363–7.
[3] A. Steyermark, *Anal. Chem.* 1948, **20**, 368–70.

heated to boiling, and if a solution results the same procedure is used in the estimation. When the material is still wholly or partially insoluble a sample is crushed in an agate mortar and a little of the finely divided powder is heated with more phenol (usually 6–8 crystals) and propionic anhydride (6–8 drops) until a solution is obtained. A few substances (e.g. pentamethyl-anisole) must be heated in a sealed tube at 135° C. for 2 hr. as they distil out of the reaction mixture and solidify in the condenser in the normal apparatus. The tube is cooled, opened, the mixture transferred to the reaction flask and the alkyl halide removed by distillation in the usual way.

Solids which dissolve easily in the phenol-propionic anhydride mixture are weighed roughly into a small tinfoil cup (made by pressing a disk 1·5 cm. in diameter over the rounded end of a glass rod 5 mm. in diameter) which is closed by pinching the open end between the thumb and two fingers and after standing for 1 min. is weighed accurately. Solids which dissolve only with difficulty are weighed into the reaction flask and the tinfoil added separately.[1] Liquids and hygroscopic solids are weighed into a small glass cup (4 mm. in length and 4 mm. in diameter) fitted with a glass stopper, or into a gelatine capsule, and tinfoil is added. The reaction mixture does not bump in the presence of stannous iodide, and the addition of pieces of broken porcelain is unnecessary. The weight of tinfoil added must never exceed 20 mg., and if possible not more than 12 mg. is added, as a large reduction in the hydriodic acid concentration causes low values.

1. *Volumetric method*

The delivery tube, spiral and receiver are washed with dis-tilled water, the water seal is made at the top of the tube, the trap is charged (1 ml., to about one-third its volume) with a sus-pension of red phosphorus in a solution of cadmium sulphate (5 %) through the filling tube which is then closed with a cork. A solution of sodium acetate in glacial acetic acid (2 ml., 10 %) and bromine (5 drops, free from iodine) are measured into the receiver which is slipped over the delivery tube and

---

[1] Cf. J. J. Chinoy, *loc. cit.* (p. 210, n. 3).

spiral by moving the apparatus outwards about the horizontal tube clamped loosely in a split cork. A small plug of cotton wool, moistened with formic acid, is placed at the mouth of the exit tube to stop bromine vapour from passing into the laboratory.

The sample is transferred to the reaction flask by platinum-tipped forceps, phenol (five crystals for substances which dissolve easily and eight crystals for compounds which dissolve only with difficulty, A.R.), propionic anhydride (8 drops) and hydriodic acid (2 ml., substances having resistant alkoxyl groups and natural products of unknown composition require a considerable excess, see below) are added. The joint of the reflux condenser is moistened with hydriodic acid, the reaction flask is attached immediately and water is run through the condenser. When the preliminary test has shown it to be necessary the sample, phenol and propionic anhydride are warmed carefully over a small micro-burner flame until the material dissolves, the mixture is cooled and then hydriodic acid is added. The carbon dioxide generator is connected to the apparatus and a slow stream of gas is bubbled through the solution so that there are never more than two bubbles rising through the liquid in the receiver at the same time (see below). The reaction mixture is heated gently with a micro-burner, fitted with a chimney, during 30 min. for a methoxyl group and 40 min. for an ethoxyl group after which time the reaction should be finished. It is advisable whenever possible to do a duplicate determination and heat the mixture for half as long again. The cork holding the apparatus is raised in its clamp until the end of the delivery tube is about 2 cm. above the liquid in the absorption tube, the burner is removed and the cork is taken out of the water seal.

The delivery tube is washed well, inside and outside, with distilled water, and the contents of the receiver, together with the washings, are transferred quantitatively into a conical flask fitted with a ground-glass stopper and containing an aqueous solution of sodium acetate (5 ml., 20%). A tap at the bottom of the receiver simplifies this operation.[1] Formic acid (2 drops) is added down the wall of the flask and mixed with the

[1] Cf. A. Elek, loc. cit. (p. 210, n. 3).

liquid before more (2 or 3 drops) is added to destroy the bromine. This is shown when a very small drop of methyl red is not decolorized. Potassium iodide solution (2 ml., 10%) and sulphuric acid (5 ml., 2N) are added, the stopper is placed in the flask, and after standing for 2 min. the iodine is titrated rapidly with sodium thiosulphate (0·02N, see below) using starch solution (4–6 drops, 1%) as the indicator.

A blank determination must be carried out on the reagents.

Percentage of alkoxyl

$$= \frac{\text{ml. of sodium thiosulphate} \times \text{N} \times \text{mol. wt. alkoxyl group} \times 100}{\text{weight of sample} \times 6000}.$$

Accuracy ± 1%.

PREPARATIONS: *Sodium thiosulphate.* Sodium thiosulphate (50 ml., 0·1N) is measured into a graduated flask (250 ml.), amyl alcohol (1·5 ml.) is added and the solution made up to the mark. The thiosulphate is standardized against iodine as follows. A standard solution of potassium dichromate (7 ml., 0·05N) is measured from a microburette into a conical flask fitted with a ground-glass stopper, water (20 ml.) and potassium iodide solution (2 ml., 5%) are added and the mixture is acidified with sulphuric acid (5 ml., 2N). The stopper is placed in the flask, and after standing for 2 min. the iodine is titrated rapidly with the sodium thiosulphate to be standardized. Starch solution (3 drops, 1%) is used as the indicator.

Hydriodic acid. Hydriodic acid (sp.gr. 1·7), which is prepared specially for Zeisel estimations, can be bought. It must be stored away from light and air to prevent decomposition and the separation of iodine, as low values result from the use of acid with a specific gravity lower than 1·7. It has been reported recently that colourless acid is not needed.[1]

Carbon dioxide. A carbon dioxide generator, or a cylinder with a reducing valve is used. The gas is passed through a solution of sodium carbonate to wash it free from hydrochloric acid vapour. A very accurate adjustment of the rate of flow is possible with a screw clip on rubber tubing (4–6 mm. bore) with a piece of string inside (10–15 cm. long).

[1] A. Steyermark, *loc. cit.* (p. 211, n. 3).

2. *Gravimetric method*

For this method the apparatus used differs only from that described above in having a straight delivery tube and a narrow receiver (7–8 mm. diameter at the lower end and 50 mm. long, fig. 28). The delivery tube and receiver are cleaned (see above) in a chromic acid-sulphuric acid bath, washed with water and with alcohol. An alcoholic solution of silver nitrate (2 ml., see below) is measured into the receiver which is placed in position with the end of the delivery tube within 1–2 mm. of the bottom (the small bubbles leaving the end of the tube should be flattened against the receiver). The trap is filled with sodium thiosulphate and cadmium sulphate, and the reaction mixture is prepared and heated as described under the volumetric procedure.

The first signs of a precipitate should be seen at the lower end of the delivery tube after approximately 3 min., and after 10 min. there should be no apparent increase in the amount of solid formed. Heating is continued, however, so that the last traces of the alkyl halide distil over and are absorbed. The solid on the inside and outside of the delivery tube is removed by washing alternatively with distilled water acidified with nitric acid (free from halogens) and alcohol until the level of the liquid in the receiver is about the middle of the wide part of the tube. Any particle which cannot be removed by washing is taken off with a feather. Concentrated nitric acid (5 drops, halogen-free) is added to the mixture, and the receiver is placed in a gently boiling water-bath until the silver iodide coagulates and sinks. High values result if too much or too little nitric acid is added. The mixture is cooled under the tap and the solid is aspirated into a filter tube, washed, dried and weighed:

1 mg. of silver iodide $\equiv 0\cdot1321$ mg. of —OCH_3,

1 mg. of silver iodide $\equiv 0\cdot1918$ mg. of —OC_2H_5.

Accuracy $\pm 1\%$.

PREPARATIONS: *Hydriodic acid and carbon dioxide.* See above, p. 214.

Alcoholic silver nitrate solution. Silver nitrate (4 g., A.R.) dissolved in alcohol (100 g., 95 %) is boiled for 4 hr. on a water bath under a reflux condenser. The solution is left to stand for 2 days and then decanted from any silver which separates and is stored in a brown bottle.

The alkyl halide is transformed quantitatively into silver iodide only when a freshly prepared solution of silver nitrate is used.[1] A correction can be applied when a stock solution is used by adding 0·12 mg. to the weight of the final precipitate for every 2 ml. of silver nitrate measured into the receiver. After the solution has aged for 6 months the results tend to be low and a new solution should be prepared.

PROPOXYL AND BUTOXYL GROUPS

Propoxyl and butoxyl groups can be estimated by Zeisel's method, but the usual form of apparatus (see p. 209) must be changed so that propyl iodide and butyl iodide can be distilled quantitatively out of the reaction mixture. These halides are not as volatile as methyl and ethyl iodides, and an apparatus is constructed with a smaller cooling surface but which is still sufficient to condense most of the hydriodic acid and return it to the reaction flask.[2] This method is preferred to the estimation based on the dichromate oxidation of butyl alcohol.[3]

METHYLIMINO AND ETHYLIMINO GROUPS

ESTIMATION BY HYDRIODIC ACID

$$\rangle N—R+HI \longrightarrow \rangle NR.HI,$$

$$\rangle NR.HI \longrightarrow RI.$$

Methyl- and ethyl-imino groups are estimated by a method similar in principle to that used for methoxyl and ethoxyl compounds. The substance reacts with constant boiling hydriodic acid to form a quaternary ammonium salt which decomposes on heating and splits off the alkyl halide.

[1] A. Friedrich, *Z. physiol. Chem.* 1927, **163**, 141–8.
[2] B. M. Shaw, *J. Soc. Chem. Ind.* 1947, **66**, 147–9T.
[3] Cf. J. J. Levenson, *Ind. Eng. Chem.* (Anal. ed.), 1940, **12**, 332–7.

The apparatus is as shown in fig. 29.[1] The gases from the reaction mixture are not forced through condensing hydriodic acid and so no hot vapours enter the washing solution. The rate of flow is constant once it is established, and when the temperature reaches 360° C. the apparatus can be left unattended.

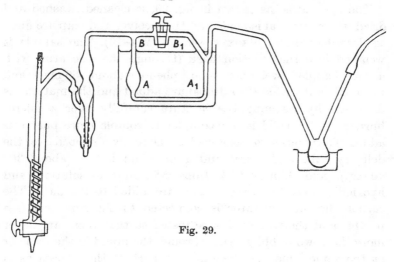

Fig. 29.

The iodide can be determined either gravimetrically or volumetrically as described under alkoxyl (see pp. 207–16). If the former method is used a suspension of red phosphorus in a solution of cadmium sulphate or a mixture of cadmium sulphate and sodium thiosulphate is placed in the trap. A suspension of red phosphorus in water does not absorb carbon dioxide and hydriodic acid completely and hydrogen sulphide passes through and may form a precipitate of silver sulphide.[2] The volumetric procedure is simpler and quicker than the gravimetric, and although traces of hydrogen sulphide do not interfere the mixed washing solution of red phosphorus in cadmium sulphate is used.

Inconsistent or low results have been reported in the analysis of the decomposition products of lactoflavin and synthetic

[1] A. Friedrich, *Mikrochemie*, 1929, **7**, 195–201.
[2] S. Edlbacher, *Z. physiol. Chem.* 1918, **101**, 278–87; A. Friedrich, *loc. cit.* (p. 216, n. 1).

flavins, but the values are accurate if these compounds are first brought into solution by warming, or boiling gently with a few crystals of phenol and a few drops of propionic anhydride.[1] To make sure that results are reliable it is advisable to dissolve each sample in this way before beginning the analysis.

The apparatus, as shown in fig. 29, is cleaned, washed and dried, the water seal is made and the receiver and trap are filled as described under alkoxyl (see pp. 211–16). The sample is weighed into the reaction flask (through the side arm) with a long-handled weighing tube, phenol (about 50 mg.) and propionic anhydride (3–5 drops) are added and the material is dissolved by warming the mixture carefully over a micro-burner. If the solid is not completely soluble more phenol is added. The reaction vessel is cooled, the receiver slipped over the delivery tube and spiral and ammonium iodide (about 30–40 mg.), gold chloride (1–2 drops, 5 % aqueous solution) and hydriodic acid (2 ml., see p. 214) are added to the flask. The carbon dioxide generator is connected to the side arm (see p. 214) and the rate of flow adjusted so that there are never more than two bubbles rising through the liquid in the receiver at the same time. The rectangular part of the apparatus is immersed in a water bath at 90° C., and with the stopcock open as shown, the flask is heated gradually (about 1 hr. to reach the temperature of 360° C.) in a metal pan filled with powdered copper oxide (residues from old combustion tubes). The temperature is recorded with a thermometer placed in the copper oxide. If the reaction mixture is heated too rapidly hydrogen sulphide and hydriodic acid are carried over with the carbon dioxide into the washing solution and they are not absorbed completely. The bath is kept at 350–360° C. for at least 30 min. The carbon dioxide is turned off and the stopcock is turned through 90° closing B_1 to B and the vertical groove then connects B with the outside air. The flask is left to cool and the hydriodic acid in the lower part of the rectangular piece of apparatus is sucked back into the reaction flask (if necessary a piece of rubber tube is slipped over the side arm and the acid is sucked back). A few drops of fresh hydriodic acid are added through the

[1] R. Kuhn and H. Roth, *Ber.* 1934, 67, 1458.

vertical opening of the stopcock and are sucked into the re-action flask. The receiver is replaced by another, freshly filled, the stopcock is turned to its original position, carbon dioxide is bubbled through the reaction mixture and the distillation is repeated at least once more. It is unnecessary to renew the washing liquid in the trap. The alkyl halide absorbed in the receivers is estimated as described on pp. 212–16.

METHYL AND ETHYL GROUPS ATTACHED TO SULPHUR

The procedure and apparatus used for the estimation of methyl and ethyl groups attached to sulphur are as described under methyl- and ethyl-imino groupings (see pp. 216–19).[1] A washing solution of sodium thiosulphate (5%) and sodium carbonate (0·5%) is used.[2] The volumetric procedure is preferred to the gravimetric, as in this estimation some hydrogen sulphide generally passes into the silver nitrate solution even when the stream of carbon dioxide is very slow and with cadmium sulphate in the washing solution (see p. 210).

METHYL GROUPS ATTACHED TO CARBON
ESTIMATION BY OXIDATION

$$R-\underset{\underset{CH_3}{|}}{CH}-CH_2-R' \longrightarrow CH_3COOH.$$

Hydrocarbon side chains are oxidized with a mixture of chromic and sulphuric acids to form an equivalent of acetic acid. This reaction is used in the estimation of side-chain methyl groups by titrating the acetic acid formed.[3] The method is of considerable importance in the elucidation of structural formulae, but care must be taken in interpreting the results, as the amount of acetic acid is not always quantitative. When the methyl group is attached directly to an aromatic ring the value is low, and

[1] F. Vieböck and C. Brecher, *Ber.* 1930, **63**, 3207–10.
[2] K. H. Slotta and G. Haberland, *ibid.* 1932, **65**, 127–9.
[3] R. Kuhn and H. Roth, *Ber.* 1933, **66**, 1274–8; R. Kuhn and F. L. Orsa, *Z. angew. Chem.* 1931, **44**, 847–53; see pp. 201–7.

when a *gem*-dimethyl group or *tert.*-butyl group is oxidized the product is acetone and not acetic acid. Ethoxyl groups can be differentiated from methoxyl groups, and if necessary the two can be estimated in a mixture as the former gives a theoretical yield of acetic acid, whereas the latter is destroyed completely. The total amount of alkoxyl can be determined as described under Ziesel's method (see pp. 207–16).

The apparatus is as shown in fig. 27 (see p. 205). The solubility of the finely powdered solid is found approximately by dissolving small samples in the oxidizing mixture. The substance (to neutralize about 5–10 ml. of 0·01N alkali) is weighed into the reaction flask, the reflux condenser is attached and the cooled oxidation mixture of chromic acid and sulphuric acid (5 ml., see below) is added. The reaction mixture is heated for ½ hr. and then left to cool, the glass rod is taken out, the condenser is rinsed into the flask with water (5–7 ml.) and the flask is disconnected from the apparatus.

Liquids and solids which form volatile products on oxidation are sealed in a capillary tube and heated in a pressure tube (about 30 cm. long) together with the oxidizing agent (5 ml.) for 1½ hr. at 120° C. The pressure tube is left to cool, opened and the contents washed into the distillation flask (see p. 212).

The excess chromic acid is reduced by adding a dilute solution of hydrazine (see below) drop by drop until the first tinge of green is seen, but hydrazine *must not* be added until a green solution results. An aqueous solution of sodium hydroxide (6 ml., 5N) is added to neutralize the acid, and the reaction mixture is just acidified with phosphoric acid (1 ml., $d = 1·7$), two or three pieces of pumice are added and the acetic acid is distilled off and estimated as described under acetyl and benzoyl groups (see pp. 203–7).

PREPARATIONS: *Chromic acid* (5N). Chromic anhydride (168 g.) is dissolved in water (1 l.), and the solution is sucked through a filter having fine pores. The oxidation mixture is made by adding sulphuric acid (5 ml., sp.gr. 1·84) to the solution (20 ml.).

Hydrazine hydrate. Hydrazine hydrate (10 ml.) is diluted with water (10 ml.).

SUBJECT INDEX

[Page numbers followed by (r) refer to particular compounds used as reagents.
Page numbers followed by (p) refer to the preparation of reagents or solutions
of reagents.]

INDEX OF AUTHORS

Treon, J. F., 195, 196
Tresider, R. S., 99
Tropsch, H., 37
Trozzolo, A. M., 145
Truffault, R., 7
Tschugaeff, L., 55
Tucker, I. W., 92
Turowa-Pollak, M. B., 5

Uchino, T., 161
Uhrig, K., 14
Underhill, E. J., 162
Urbain, M., 95
Utzinger, M., 208

Vandenberghe, H., 185
Vanderzee, C. E., 197
Vanetten, C. H., 199
Van Natta, F. J., 99
Van Slyke, D. D., 183
Veen, H. van der, 28
Veibel, S., 153, 154
Venkataraman, K., 195
Verain, M., 161
Verley, A., 42
Vieböck, F., 208, 219
Volkholz, H., 161
Voorhees, V., 2, 4, 5, 7
Vorländer, D., 161
Vortmann, G., 104
Vries, T. de, 196

Wagner, C., 158
Wagner, E. C., 131
Wagner, J., 163
Wagner, R., 153, 193
Wakeman, R. L., 14
Waldschmidt-Leitz, O., 5
Walker, F. T., 27
Wallerius, G., 185
Walters, T. M.,
Warren, G. G., 55
Warshowsky, B., 68
Waterman, H. I., 146, 155
Waters, T. M., 185
Waters, W. A., 4, 42
Watson, H. B., 130
Weatherburn, A. S., 120
Weatherburn, M. W., 120
Weber, E., 203
Weidlin, E. R., Jr., 34
Weiler, G., 6
Weisenberger, E., 204, 206

Weisz, M., 97
Weith, A. J., 87, 88, 104
Wendt, G. L., 126, 127
Werkman, C. H., 142, 166
Werner, A., 9
Wernimont, G., 51
West, E. S., 42
Westheimer, F. H., 129
Wetlaufer, L. A., 99
Weygand, C., 9
Whalley, W. L. O., 70
Wheeler, D. H., 28
White, D. L., 128
White, E. P., 210
White, E. V., 210
White, T., 208
Wichers, E., 32
Wiele, B. M., 199
Wijs, J. J. A., 19, 20, 21, 22, 23, 24
Wilcox, C. S., 51
Wild, F., 142, 156
Williams, J. G., 136, 137
Williams, K. A., 21, 22, 26
Williams, R. D., 92
Willits, C. O., 42, 191
Willstätter, R., 2, 5
Wilson, H. A. B., 211
Wise, M., 191
Wislicenus, W., 82
Witte, K., 196
Wittwer, Ch., 84
Wöhler, P., 131
Wolf, R., 157
Wolff, G., 83
Wolfrom, M. L., 209, 210
Wright, G. F., 56, 57, 210
Wright, L. O., 70

Yagoda, H., 132
Yoe, J. H., 156, 161
Young, S. W., 185
Youtz, M. A., 126
Yushkevich, S., 35

Zahn, V., 99
Zartman, W. H., 7, 8
Zaugg, H. E., 64, 66
Zeisel, S., 207, 220
Zelinsky, N. D., 5
Zerewitinoff, Th., 55, 56, 86
Zetzsche, F., 6
Ziliotto, H., 199
Zoff, A., 105